The Cambridge Manuals of Science and
Literature

PLANT-ANIMALS

The Cambridge History of Scandinavia

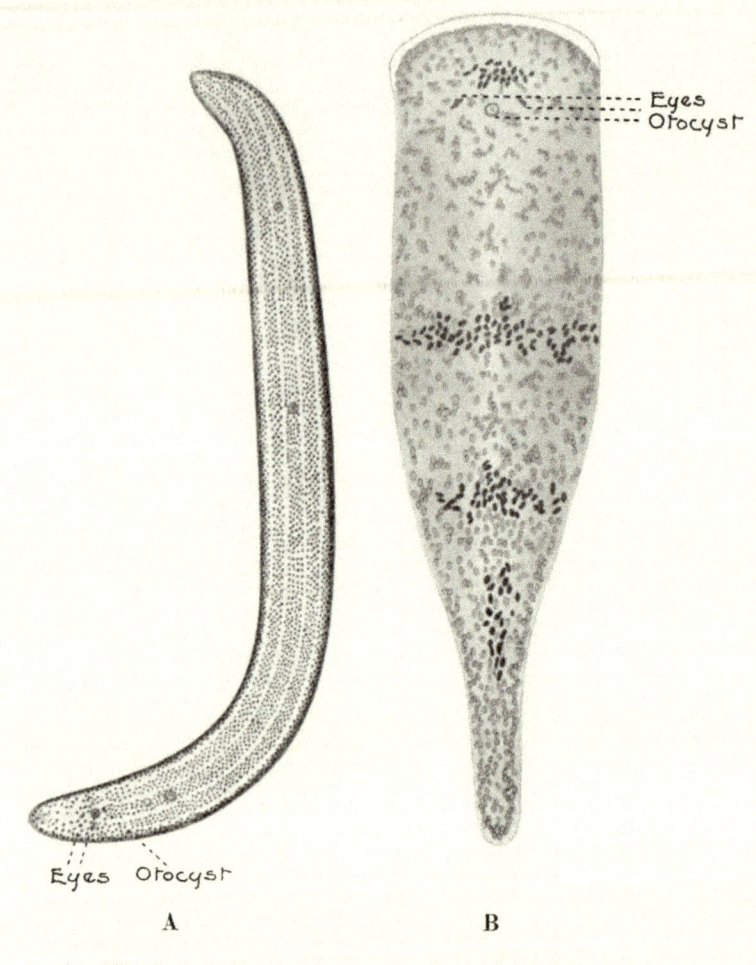

A.   The Green Plant-animal (*Convoluta roscoffensis*).
B.   The Yellow-brown Plant-animal (*Convoluta paradoxa*).
40 times natural size.

This image is available in colour for download from www.cambridge.org/9781107605893

PLANT-ANIMALS

A STUDY IN SYMBIOSIS

BY

FREDERICK KEEBLE, Sc.D.

PROFESSOR OF BOTANY IN
UNIVERSITY COLLEGE, READING

Cambridge :
at the University Press
1910

CAMBRIDGE UNIVERSITY PRESS
Cambridge, New York, Melbourne, Madrid, Cape Town,
Singapore, São Paulo, Delhi, Mexico City

Cambridge University Press
The Edinburgh Building, Cambridge CB2 8RU, UK

Published in the United States of America by Cambridge University Press, New York

www.cambridge.org
Information on this title: www.cambridge.org/9781107605893

First published 1910
First paperback edition 2012

A catalogue record for this publication is available from the British Library

ISBN 978-1-107-60589-3 Paperback

Additional resources for this publication at www.cambridge.org/9781107605893

*With the exception of the coat of arms at
the foot, the design on the title page is a
reproduction of one used by the earliest known
Cambridge printer John Siberch 1521*

PREFACE

DURING some ten years' work in a small marine laboratory in Brittany it has fallen to me not infrequently to attempt to explain to curious visitors what were my objects in going to and fro upon the shore, in wading among the sea-weeds and in bringing into the laboratory minute, worm-like animals which represented often my sole "catch."

I discovered that many of the visitors to the laboratory became interested in the work that was going on, and that, though they disclaimed a knowledge of biology, they followed with understanding and interest the story of the behaviour and life histories of "the worms":—indeed, they succeeded generally in putting to me pertinent and unanswerable questions with respect to these "plant-animals."

The pleasant recollection of hours spent in this way is responsible primarily for my undertaking to contribute this volume to *The Cambridge Manuals of Science and Literature.*

If it succeeds in interesting the layman, success will be due to the severe educational régime to which my visitors submitted me in their cross-questionings as to the bearings and objectives of my biological work.

If it fails, they must bear the blame: for had they not exhibited a fondness for "Convoluta" I should scarcely have ventured to publish its doings to the world at large Of these friends I would mention particularly Mr Alfred Dutens, whose interest in "Convoluta roscoffensis" has been a source of constant encouragement to me.

The biological facts recorded in this volume are the outcome of researches carried on for some years by Professor Gamble and myself, and, subsequently, without Professor Gamble's co-operation.

Throughout the whole time during which the work has been in progress, it has benefited more than may be stated explicitly by the unremitting assistance rendered by my wife. To her, are due the long and patient records of the periodic changes of behaviour of the plant-animals—*Convoluta roscoffensis* and *C. paradoxa*:—records which entailed visits to the Convoluta colonies at all phases of the tide and at all hours of the day and night. Though an adequate expression of my thanks to my wife were out of place here, I beg leave to give myself the pleasure of acknowledging how great has been her share in this work.

The original memoirs, giving detailed accounts of the life histories of the plant-animals, have appeared in the *Quarterly Journal of Microscopic Science*. A list of these memoirs and of other researches to which reference is made in the text is included in the short bibliography appended to this volume. The dates, enclosed within brackets in sundry places in the text, refer each, to the year of publication of the research which is cited and indicate that the title of the re-search in question may be found in the bibliography under that date.

The black and white illustrations have been pre-pared specially for this volume by Mrs Seward from the original drawings made by Miss Dorothea Richardson in the laboratory at Trégastel. I am deeply indebted to Mrs Seward and Miss Richardson for their kind assistance, and to the skill and patience which they have bestowed on the drawings I offer a sincere and admiring tribute.

Should the reader find that the main arguments exposed in the course of the volume are intelligible, he may, perhaps, be inclined to forgive the use, which I hope is as occasional as inadvertent, of unfamiliar biological terms. I have endeavoured to avoid this pit-fall, but have doubts as to the completeness of my success. I shall be obliged therefore if readers will point out passages which require elucidation, so that, in the event of another edition being published, the defects may be remedied.

FREDERICK KEEBLE.

Trégastel,
    Côtes-du-Nord,
        France.
            September, 1910.

CONTENTS

PART I

THE BEHAVIOUR OF THE PLANT-ANIMALS

CHAP.   PAGE

I. Introductory : the worms, Convoluta roscoffensis and Convoluta paradoxa : their habits and habitats 3

II. The origin and significance of the habits of Convoluta roscoffensis and Convoluta paradoxa . 37

PART II

THE NATURE OF THE PLANT-ANIMALS

III. The green cells of Convoluta roscoffensis and the part they play in the economy of the plant-animal . 75

IV. The origin and nature of the green cells of Convoluta roscoffensis 100

V. The significance of the relation between coloured cell- and animal-constituents of the plant-animals 130

BIBLIOGRAPHY 159

INDEX 161

WITH TEXT-FIGURES, 1—22.

COLOURED FRONTISPIECE. The Green Plant-Animal, Convoluta roscoffensis and the Brown Plant-Animal, Convoluta paradoxa.

PART I

THE BEHAVIOUR OF THE PLANT-ANIMALS

CHAPTER I

BIOLOGISTS who devote themselves to the investigation of the life histories and life processes of the
lower animals are apt to encounter the criticism:
why expend pain and labour on insignificant creatures
when so much remains to discover with respect to
the higher animals, including man himself?

This perfectly legitimate criticism admits of a con-
clusive reply and, since it is possible that a question
of the kind may arise in the mind of anyone taking
up this book, it shall be answered forthwith.   The
reply may take one of three forms.   In the first place,
it may be urged that the most important modern
biological discoveries have resulted from researches
into the life histories of the lower organisms.   Modern
surgery relies for much of its technique on the results
of investigations into the physiology of the bacteria.
Yet more recently, the experimental elucidation of
the life-histories of the protozoa—the lowest group of

1—2

animals—has laid the foundation of a great and increasing body of knowledge with respect to the cause of malaria, sleeping sickness, and other tropical diseases.

In the second place, it may be urged that, the more complex the organism, the more difficult it is to use the results of observations upon it for the purpose of generalising on important biological problems such as those of the origin of instinct and habit, or of the meaning of heredity and the course of evolution. The higher the organism, the more it has covered up the tracks along which the species to which it belongs has travelled. For this reason alone, the study of the lower organisms is not only to be justified but also urged on zoologists as one bound to lead to results of the greatest value.

In the third place, it has yet to be proved that the higher animals differ in any fundamental respect from more lowly forms of life. Hence, if, as a physiologist must hold, such differences as exist between higher and lower forms are differences of degree and not of kind, it follows that an increased knowledge of the nature of the lower organisms connotes also an increase in knowledge with respect to the higher organisms.

On these grounds, the patient and exhaustive study of the lower organisms is to be justified. Nay more, if the reasons for this study are valid they

should serve to induce some of the younger generation of physiologists to devote their attention to a
field of research both rich in promise and too little
cultivated by the men of science of this country.

Though the results recorded in this volume are
but modest, throwing here and there only a faint
light on the problems which they raise, nevertheless
they suffice to demonstrate that more skilful observers
would, by taking up similar subjects of investigation,
make notable contributions to the science of comparative physiology.

Having vindicated the importance of research on
the lower organisms, let us proceed to our task.

The plant-animals whose life histories and habits
form the subject of this volume are two simple,
marine worms, *Convoluta roscoffensis* and *Convoluta
paradoxa* (Frontispiece). Both are small, though large
enough to be seen easily by the unaided eye, and both
are conspicuous by reason of their colours. C. roscoffensis is dark, spinach green, and C. paradoxa
yellow-brown.

Even among worms they occupy a lowly place.
Unlike the higher members of this group, C. roscoffensis and C. paradoxa are unsegmented. Instead
of consisting, like garden worms, of a series of ring-
like pieces, the bodies of our plant-animals are in one
piece and, consequently, bear no ring-like markings

S.

I.

S.

II.

Fig. 1. The distribution of the colonies of Convoluta roscoffensis on
the sea-shore. I. at spring-tidal periods (low water): II. at
neap-tidal periods (low water). Though a colony remains fixed
in position, its size waxes with the spring tides and wanes with
the neap tides. C, C = the colonies. S. = sea

on their surfaces (Frontispiece). Imagine a minute, elongated fragment of a most delicate leaf, some $\frac{1}{8}$ in. long by $\frac{1}{16}$ in. broad, and you have a picture of C. roscoffensis. Imagine, further, myriads of such green, filmy fragments lying motionless on moist, glistening patches of a sunny beach between tide-marks and you see the species in its native habitat (Fig. 1). To find C. paradoxa at home it is necessary to follow the receding tide, to gather handfuls of the brown seaweeds (Fig. 2) which are exposed towards the low-water limit of the larger tides and to allow the tips of the weeds to dip into water in a white dish. Singly from their hiding-places chubby, brown C. paradoxa come gliding down with rounded "head" and pointed "tail" to swim uneasily in the water of the dish. C. roscoffensis is pre-eminently gregarious, C. paradoxa by comparison is solitary. Sand from a Convoluta patch scooped up in a cup contains many thousands of C. roscoffensis; a patient fishing throughout the time of low tide may result in a catch of fifty, or at most a hundred, specimens of C. paradoxa.

The surface of the bodies of the plant-animals is somewhat slimy; particularly in C. roscoffensis, and is covered by fine cilia (Fig. 3) which, during the life of the animals, are in constant motion. The cilia, which are protoplasmic projections from the super-ficial cells, serve, by their unceasing movements, to row the animal through the water.

C. paradoxa possesses, in addition to cilia, occasional, stouter, bristle-like structures which stick out
from its body, chiefly in the "tail" region (Fig. 16,
p. 84). These structures serve, when put in action
by the animal, to pin it down and thus enable it to
stop and stick in any position.

Fig. 2.   Convoluta paradoxa (C) attached to sea-weeds
of the paradoxa zone. (Magnified eight times.)

In both animals, the sides of the body are flexed
beneath the under surface, and together form a groove
which, in C. paradoxa, serves to fit the animal saddle-
wise to the fine sea-weeds over which it glides (Fig. 2).
This animal, in its general progress, appears almost
to flow over the substratum on which it is moving.

Occasionally, however, on meeting with an obstacle it rears its head-end, caterpillar-wise, relaxes the grip of its flexed sides, readjusts them to the surface and glides on with stealthy motion.

Though we have called C. roscoffensis and C. paradoxa simple worms, it is not to be inferred that the structure of their bodies is really simple. Both species possess a well-defined nervous system and efficient sense-organs. At the front or "head" end of the body, on the upper surface, a little way behind the anterior end, lie two eyes right and left of the median line (Frontispiece and Fig. 3). Though of the simplest construction, each consisting of a minute spot of orange pigment lying over nervous tissue, the eyes are efficient for distinguishing light of different intensities. Numerous orange-pigmented glands, scattered over the surface of the body, function probably as accessory eyes. Between the two eyes, in the median line on the dorsal (upper) side of the body of either species, lies the otocyst (Frontispiece and Fig. 3, *OT*). It consists of a hollow sphere of nervous tissue enclosing a space within which lies a small lump of chalk.

Like a pea in a thimble, the heavy, chalky mass, or otolith, lies freely in the otocyst, and, if the position of the animal change with respect to the line of action of gravity,—the vertical—the otolith falls or rolls on a new part of the otocyst-wall. Pressing on this area it acts as a stimulus to the nervous tissues beneath. As the result of stimulation of this tissue, nervous impulses may be despatched to the muscles of the body, and, causing them to contract, give rise to movements of the body which are definite in direction.

Thus the otocyst serves as an indicator of the line of gravity; in other words it acts as the organ

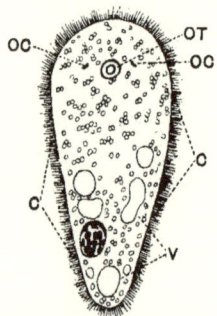

Fig. 3. Young Convoluta paradoxa. C = cilia covering the surface of the body. OT = otocyst. OC = eyes. V = empty digestive vacuoles.

for gravi-perception. By its means, the animal is able to orientate itself with respect to the vertical, and so to find its way downward or upward.

That the otocyst does indeed serve this end has been established by experiments with other animals, and may be inferred in the case of C. roscoffensis from the following facts. Occasionally, among just-hatched larvæ specimens occur which fail to respond like their fellows to gravitational stimulus. Such specimens are found, on microscopic examination, to lack properly developed otocysts. For example, if numbers of C. roscoffensis larvæ are taken up with water into a glass tube and the tube is shaken slightly, the animals come down, some tumbling, some curvetting. These animals in general respond to vibration by a geotactic movement—that is, one having reference to the line of action of gravity—but the one or two, devoid of otocysts, fail to descend, remain glued to the side of the tube and are dislodged with the greatest difficulty.

As indicated already, the bodies of Convoluta possess a well-developed system of muscles by the ordered contractions of which the movements of the animals are effected.

The digestive system is of a primitive order. A well-developed mouth, capable of a wide gape, occurs on the under side of the body rather nearer the "head" than the "tail" end. The mouth communicates by a short gullet, not with a distinct digestive tube, but with a loose, central tissue. Hence food which is ingested passes through the mouth to the gullet whence it is distributed to improvised spaces or vacuoles in the tissues (Fig. 3). In these vacuoles it is digested. The undigested residue is discharged at any point of the body, generally, however, toward the hinder end.

Neither species of Convoluta possesses a circulatory system. In the absence of heart and blood-vessels, the distribution of the nutritive substances derived from the food is effected in a primitive manner, the materials being passed from cell to cell.

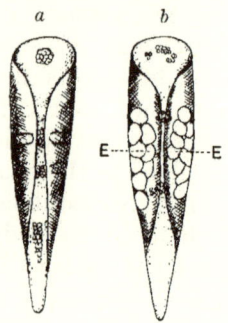

Fig. 4. Convoluta paradoxa. *a*. Seen from ventral surface, showing the folds of the sides of the body. *b*. An animal with nearly ripe eggs (E).

There is, moreover, no excretory apparatus, and the waste products are not discharged from the body but remain and accumulate in the tissues.

Both C. roscoffensis and C. paradoxa are hermaphrodite, each animal possessing male and female reproductive organs, the essentials of which are, respectively, spermatozoa and egg-cells. The eggs are numerous and attain to so considerable a size that

they may be seen lying in rows in the bodies
of "ripe females," that 'is, animals in the female
stage (Fig. 4, *b*, *E*). The eggs are fertilized in the body,
though the spermatozoa which effect fertilization
are derived from another individual of the same
species. After fertilization, the eggs are discharged
in groups or clutches of from about eight to fifteen
or more. As it is extruded from the body the
egg-clutch becomes surrounded by a transparent,
mucilaginous, sticky capsule secreted by the glands
on the surface of the skin. A clutch of eggs of

Fig. 5. Egg-capsule of Convoluta paradoxa. Each egg is contained
in an egg-membrane and the group of eggs is enclosed by a
common capsule. (Magnified twenty times.)

C. roscoffensis is recognisable to the trained eye as
a minute, more or less transparent sphere of about
the size of a small pin's head. The egg-clutch of
C. paradoxa is of a similar size ; but, owing to the
presence of pigmented granules, it is of a rufous
colour (Fig. 5).

C. roscoffensis lays its eggs on the beach just
beneath the surface of the sand : C. paradoxa deposits
them on the fine sea-weed lower down the shore.

The habitat of C. roscoffensis is restricted and localised (Fig. 1). This gregarious species occurs within a well-defined zone of the foreshore of sandy beaches of Normandy and Brittany. Elsewhere it is unknown.

An observer, walking at low tide seaward across a golden beach in Brittany, passes scattered granite rocks scantily clad with yellow-brown patches of seaweed adventuring landward and before he reaches the main belt of brown seaweeds, some yards landward of the thin line of green Cladophora which lies bleaching in the sun, he may see dark, spinach-green glistening patches—the colonies of C. roscoffensis. He must tread softly lest the patches melt away at his approach. The colonies may extend for many yards as dark green, irregular strips running more or less parallel with the shore-line, or they may consist of apparently disconnected patches varying in size from an inch or so to a yard or more across. From the intervals between the colonies, the animals are not absent. Though they are not to be seen, they may be smelt. Sand from a part of this roscoffensis zone where no animals are visible, when squeezed between the fingers, emits from the crushed, occasional Convolutas contained in it a pungent and evil smell. The odour, which is like that of decaying fish, is due to the volatile trimethyl amine which is produced by the animal.

On a peaceful beach, in quiet times, when storms and tourists are absent, the colonial patches of C. roscoffensis keep their respective outlines with surprising constancy. Day after day the several patches may be recognised, waxing in size with the spring tides, waning with the neap or slack tides (Fig. 1): larger, also, on any day soon after the tide has receded from their borders; smaller, just before the rising tide invades them. At certain times, the multitude of individuals which make up a patch may be seen lying lethargic and motionless, bathed in the sunlight and the film-like stream of drainage sea-water which oozes from the sand and flows over them seaward. On days of bright sunshine, in particular, the animals lie very still; on duller days, a constant gliding too and fro of these minute films of living matter is to be observed within the confines of a colony. It is on such occasions that the observer must tread softly, for C. roscoffensis is so sensitive to vibration that his heavy, approaching tread may send it to earth with lightning speed. How quickly the animals may make their descent from the surface may be judged from the illustration (Fig. 6) which depicts two photographs of a colony, the second taken at an interval of five minutes after the first. Three gentle taps on the sand, after the first photograph was taken, served as the signal for retreat. At that signal, the army, many millions strong, vanished with amazing swiftness and took cover underground. Lest the words "many millions" should seem to

savour of exaggeration, it may be said that one colony
of moderate size—extending over some two square
yards—was found by estimation to contain 5,600
million individuals. Of such flaky thinness are these
animals that as many as 28,000 may be packed in a

I.

II.

Fig. 6.   Response of C. roscoffensis to vibration.   Reproductions of
photographs of a colony.   I. before,   II. five minutes after the
sand had been tapped lightly with the foot   The dark patches
in I. represent vast numbers of the animals which in II. have
disappeared almost entirely below the surface of the sand.

space measuring one cubic centimetre. A search on
dark nights at low tide in the roscoffensis zone fails to
reveal any of the animals upon the surface. In such
circumstances they remain just beneath the sand.
On moonlight nights, some, but not many, may be

seen, by the light of a lantern, lying in the river-films
of their diurnal stations.

Except for a rich micro-flora and -fauna of diatoms,
bacteria and infusoria, except for a rare, solitary
enemy—another worm, a species of Plagiostoma, which
shovels live Convolutas by the hundred into its capacious body—except for an occasional, small shore-crab,
picking its way with rolling but deliberate gait over
the patches, C. roscoffensis enjoys undisputed possession of its tract of foreshore. Though the wastage
from each colony must be prodigious, every incoming
tide taking toll, yet the species, fecund and resource-
ful, rises superior to the circumstances of its environment and maintains itself in the strange situation
which fate has chosen for it.

The roscoffensis zone (Figs. 1 and 7) is as localised
as the range of distribution of the species is restricted.
The upper limit of the zone is marked by the level
reached by high water at the slackest of the neap tides:
for, further landward, C. roscoffensis could not obtain
at all tidal periods the diurnal plunge-bath without
which it does not thrive. Risk of desiccation bars
its more landward advance. The lower limit of the
zone is but a few yards seaward, for C. roscoffensis
loves the light and ensues it.

At every making tide, this zone is submerged and
C. roscoffensis becomes a submarine plant-animal
sheltering beneath the surface of the sand out of

reach of the shock of the waves. At every falling tide, as the receding waters lay bare the zone, C. roscoffensis, rises to the surface of the sand and becomes a land plant-animal, or rather, a sedentary denizen of the filmy rivers which have their sources in the sand flooded by water when the tide was full. Where the springs of drainage-water reach the surface and become rivulets cutting seaward courses, is the upper limit of the C. roscoffensis zone. Thus the colonies are so situated on the beach that they are bathed continuously in running water and receive *the maximum of light-exposure* during low water at all tidal periods. Records kept during a lunar month show that the time of exposure during low tides is very fairly constant. The time during which C. roscoffensis lies on the surface is, on the average, five and a half hours, and ranges from four and a half to six hours. Twice during twenty-four hours the roscoffensis zone is submerged and the animals live a life of darkness underground: twice the zone is uncovered and the animals are free to rise to the surface of the sand (Fig. 7). By fixing its station and adjusting its habits, C. roscoffensis succeeds to a remarkable degree in simplifying its environmental conditions. In that station, periods of inundation succeed periods of exposure at fairly regular intervals, and, by synchronising its rhythmic movements up to the surface and down below the surface with the movements of the tides,

C. roscoffensis adjusts its working days to the rhythmic changes of its environment. How remarkable is the rhythmic movement up and down we shall presently discover.

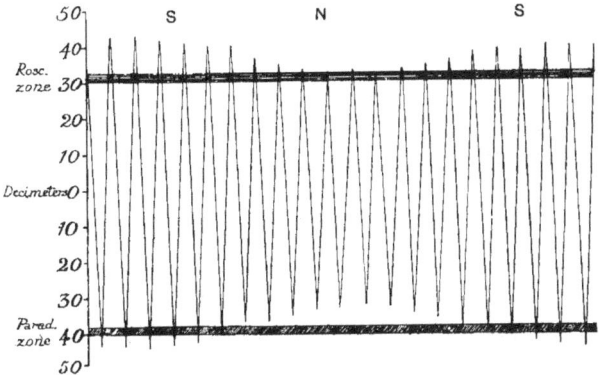

Fig. 7. Habitats of C. roscoffensis and C. paradoxa shown in relation with the rise and fall of the tides during a lunar period. S = spring tides. N = neap tides. Rosc. zone = habitat of C. roscoffensis. Parad. zone = habitat of C. paradoxa. The position of colonies of C. roscoffensis is just below the high-water level of the slackest tides. The habitat of C. paradoxa is uncovered at low water except during the slacker neap tides.

Leaving the C. roscoffensis zone and passing the rank, brown sea-weeds left high and dry by the tides, the observer paddles into the shallow water, or, if the tide is a big one, walks almost dry shod and sees the long, yellow bands of another sea-weed (Ascophyllum) swaying beneath the water of the pools or lying

2—2

prone on the soft, grey ooze of the sea-floor. The extremities of the Ascophyllum are clothed with tufts of fine, epiphytic brown and red sea-weeds. Further out, as the tide continues to fall, the browner weeds are becoming uncovered; first, the dichotomous straps of Himanthalia which spring from button or saucer-like stalks attached to the rocks, and then the finger-like Pycnophycus (Fig. 2) which extends beyond the seaward limits of even the biggest spring tides. It is among the fine weeds attached to Pycnophycus that C. paradoxa is to be found. On dangling these weeds in water, the animals come out, but as single spies not in battalions like C. roscoffensis which lies in swarms thirty yards further up the beach. The abode of C. paradoxa is less circumscribed than that of C. roscoffensis and shifts with the tides. At the onset of the spring tides, minute specimens may be taken from among the epiphytic weeds attached to the most landward of the brown sea-weeds (Fucus). During subsequent spring tides, the animals must be sought lower down the beach in the zone occupied by Himanthalia and Ascophyllum ; whilst, yet later in the same series of springs, C. paradoxa is to be found only in the Pycnophycus zone. Just where that dark brown weed ceases to be exposed at low water of the largest spring tides is the further limit of the paradoxa zone (Fig. 7). Like the Greek sailors described in *Eothen* C. paradoxa hugs the shore. Exposed now to the violence of the sea and now to the hot sun striking on the drying, emerged rocks and weeds, C. paradoxa has chosen its abiding place. But, unlike C. roscoffensis, C. paradoxa fails to finds in its station a regular recurrence of change, and hence it is constrained to shift its station during the lunar periods. At times of slack tide, the seaward part of the C. paradoxa zone is submerged continuously and the light which reaches the animals clinging to weed some feet below the surface is too feeble for their requirements. Hence, during such tides, C. paradoxa edges up landwards to the shallower water and reaches so far as the Fucus zone. During the spring tides, this latter zone is left high and dry for hours and hapless C. paradoxa stranded there would suffer from the intense isolation and also run the risk of desiccation. So, as the tides increase, it works its way down the beach, reaching, at the median spring tides, to the more seaward weeds, and at the largest springs, when these weeds may no longer harbour it in submerged peace, it treks again yet further toward the sea and takes up its station among the tangle of fine weeds which hang in tassels from the finger-like, dark brown Pycnophycus. During the slack periods, at low water, when the landward part of the Pycnophycus zone is uncovered, C. paradoxa creeps into the deepest recesses of the matted, emerged weeds. Soon, the making tide covers the Pycnophycus with an increasing load of water and C. paradoxa, clinging painfully to the floating, swaying weed, finds itself exposed to a light intensity none too high for its requirements.

Unlike C. paradoxa which, as we see, migrates periodically, its flittings coinciding with the phases of the moon, C. roscoffensis, having selected its station on the beach, maintains it in spite of time and tide. Small wonder therefore that the latter organism has learned to respond so swiftly to vibrations that it sinks below the sand at the approach of heavy feet. How sure and swift are the uprisings and downlyings of C. roscoffensis may be learned by standing at the water's edge near by the situations known to be occupied by C. roscoffensis colonies. Scarcely has the tide run off them when a faint green discolouration of the sand marks the contours of each colony, and before the water has receded more than a few yards the dark greenness of the patch indicates that all the animals have risen to the surface. Or if, when the sea is smooth, we watch the incoming tide making its way with gentlest approach toward the patches, we see the animals inert and lying massed together, bound into scum-like lumps by the muci-laginous excretion of their bodies. They lie motion-less, oblivious of the lapping waves a yard or so away. Then, as the latest wave washes over the patch, lethargy gives place to action and, in an instant,

C. roscoffensis is gone. On stormy days, when the making tide announces its landward progress angrily —thundering like ramping clouds of warrior horse— the reverberations of the sand send signals to the colonies which make their dispositions underground long before the breaking waves can reach or damage them. All these ordered goings and comings may the observer see on any day on any beach in Brittany. But, to discover more precisely the physiological methods of these purposeful movements, the laboratory must take the place of the beach, and simple scientific methods must supplement bare observation. In this way, it is possible to refer movements, so purposeful as to suggest volition, to simple, non-conscious, nervous responses to one or more of several stimuli, the chief of which are gravity and light.

Before, however, we investigate the living animals in the laboratory we may note yet another example of rhythmic behaviour in our plant-animals.

However carefully the observer seeks at low water among the exposèd weeds of the paradoxa zone, he will find no animals bearing ripe eggs. As the tides become large enough to permit of approach on foot to that zone, the animals which he obtains are, for the most part, minute, immature specimens. On succeeding days, the catch consists of larger animals, till, during the latest spring tides, it is composed chiefly of adults, many of which may contain unripe eggs. Then comes a period of slack tides when the paradoxa zone is constantly submerged beneath ten feet or more of water. At the succeeding spring tides, the same sequence of immature, young and adult animals is obtained by the collector. The absence of mature females and of deposited egg-capsules is not to be explained by a migration of gravid females to some other place more convenient for the purpose of egg-laying ; for, now and again, a solitary capsule may be found during the latest spring tides glued to the weed of the paradoxa zone. By hatching experiments carried out in the laboratory, it may be demonstrated that the time of maturing of the animals coincides with a definite tidal period. It takes either a month or a fortnight for the animals to become mature. They reach maturity at neap-tidal periods. At the beginning of these periods, or soon after, when the zone is submerged continuously for some seven or eight days, C. paradoxa lays its eggs. No matter how the conditions are altered in artificial hatching experiments, C. paradoxa is faithful to its habit : as indicated by the diagram (Fig. 8), which records the results of such experiments, the females lay their eggs only during the neap tides.

Nor is it without significance that the large yellow-brown eggs of C. paradoxa, rich in food-yolk, hatch with extraordinary rapidity. Within twenty-four to forty-eight hours of the time of laying, the larvæ, after circling actively within the capsule, burst the walls thereof and escape. Thus they have, during the remainder of the neap-tidal period, some days of comparative tranquillity and uniformity of conditions. Not for some days yet will they be exposed to the full to the chances and changes which must beset their adult lives. They are born as submarine animals, and in their earliest days are spared to some extent the buffetings which shall be theirs when, with the advent of the spring tides, they are, now, clinging to fragile weed dashed against

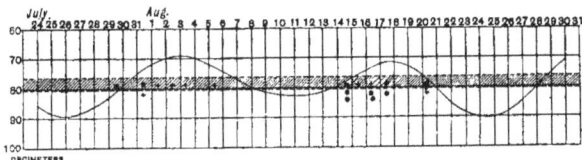

Fig. 8. Periodicity of egg-laying and hatching of C. paradoxa. The shaded band shows the position of the Paradoxa zone with respect to low water-marks of spring and neap tides. The undulating line, joining up the low water-marks of successive day-tides, is obtained by marking off, along the verticals indicating successive days, from a zero line above, the amount of vertical descent (in decimeters) of each day's tide. On those days when the undulating line falls below the shaded band, the Paradoxa zone is uncovered during low water; on those days when the low water-line lies above the shaded band, the zone is continuously submerged. The dots represent egg-capsules, the crosses signify larvæ hatched; the positions of dots and crosses give the dates on which the capsules were laid and larvæ emerged.

the rocks and, now, still clinging to the weed, left stranded, prone upon the ooze beneath the glare of an August sun.

With C. roscoffensis, also, egg-laying is a periodic phenomenon, though, in this species, the times at which it occurs coincide not with the beginnings of neap tides but with the onset of the springs. A colony of C. roscoffensis is indeed a well-drilled army. Not only do all its members take cover as one unit at a given signal, not only do the individuals keep their ranks when the order comes to climb to the surface once again, but they are born together, grow up together, mature at the same time and lay their eggs simultaneously. As a consequence, it is easy to obtain large numbers of egg-capsules, though only at definite tidal periods. To secure them, all that is necessary is to visit a fertile colony at low water during one of the earlier spring tides, tap with the foot and thus drive C. roscoffensis below the surface, scoop up a little sand, shake it with sea-water in a glass tube, and isolate the slow-sinking, transparent capsules. It is still easier, however, to rear them in the laboratory. By collecting a cupful of sand and C. roscoffensis just before the onset of a series of spring tides, bringing the cup into the laboratory, adding a little sea-water, leaving it till the plant-animals have collected on the surface, scooping them off with a watch-glass and putting them with sea-water in a large glass vessel, hundreds of egg-capsules may be obtained within a few days. The laying continues for a week or more and then, when the time of the slack tides arrives, it ceases, even though some of the animals are yet carrying mature eggs. After a barren fortnight, egg-laying begins again. Both the animals which failed to bear and those which produced eggs contribute to the fortnightly crop. The mode of egg-laying of C. roscoffensis is in some respects peculiar. Occasionally, the eggs are discharged separately one or two at a time ; but more often they are contained, as has been stated, in a common, gelatinous capsule. It happens frequently that oviposition results in a rupture of the tissues of the parent. The body becomes torn and may even break across the middle. The anterior end crawls away and, behaving like an intact animal, heals its wounds, regenerates its lost parts and recovers completely. The tail end remains near the egg-capsule, and exhibits ceaseless, revolving, "circus" movements, swimming in devious spirals ; then it comes to rest and finally disintegrates. Unlike C. paradoxa, C. roscoffensis is slow in hatching. After about four days, the larvæ begin to revolve actively within the capsule-membrane, then at the fifth to the seventh day after the eggs were laid, the egg-membranes split equatorially and the society of larvæ is set free to creep and swim within the common capsule. Suddenly they leave it, passing with ease through the mucilaginous wall, though they not infrequently return now and again to the capsule after enjoying a short spell of activity—a fact the significance of which we shall have occasion to comment upon later.

In seeking an explanation of the significance of the fortnightly periodicity in egg-laying, it is easy to conclude that the periods chosen by C. roscoffensis are the most convenient for this purpose. For in summer, to which period of the year these observations apply, low water of spring tides occurs about midday and midnight. Now, as we have learned, when the roscoffensis zone is uncovered during the darkness of night the animals do not remain on the surface. Hence, during the spring tides, C. roscoffensis has an uninterrupted period of some eighteen hours in which to lay its eggs. At other tidal periods, its leisure would be less, for, as the tide runs off the patch, the animal must come up to the light and it must remain up till the returning tide gives the signal that its vigil is at an end. In short, whereas, during the neap-tides, at which periods low water occurs in early morning and late afternoon, C. roscoffensis has two up- and two down-periods, during the springs it has only one period of compulsory "upness" in each twenty-four hours.

The weak point, however, of all such teleological

explanations is that they tend to exercise the ingenuity of men of science rather than to advance our knowledge of physiology. To which it may be answered that adaptation is as much a property of protoplasm as weight is a property of matter, and that the biologist is performing a service in showing how deep-bitten into the organism are the adaptations whereby it adjusts itself to its environment.

To this the critic replies : This is very true, but to rest content with a teleological explanation, to say that this animal does such and such a thing because it is convenient or useful for it to do that thing is to renounce profound investigation. Before this can be regarded as the proper philosophical attitude toward life, the resources of chemistry and physics must be exhausted, and the behaviour under consideration must at least be proved *not* to be due to a chemical or physical change induced by some factor or factors of the environment.

In other words, the least the physiologist can do is to attempt to discover how the adaptive trick is performed by the animal which exhibits it.

An admirable example of an apparently adaptive character, which is capable of a simple physical explanation, is given by Loeb (1909) in his brilliant essay on the influence of environment on animals.

The two species of Salamander, Salamandra atra and S. maculosa occupy distinct stations. The former species occurs in dry alpine regions of relatively low temperature ; the latter, in lower regions with plenty of water and of higher temperature.

In the dry, alpine regions S. atra deposits eggs which hatch out as land-animals; in the wet lowlands, the eggs laid by S. maculosa contain embryos in a less advanced stage of development. The young, when born, are gill-bearing and complete their development whilst leading an aquatic mode of life. Thus each species is adapted to the physical conditions of its environment.

But it has been shown that if S. atra is exposed to lowland conditions, that is, to a moist atmosphere and a relatively high temperature, it lays its eggs earlier, the young hatch out in the gill-bearing stage and development is completed during their life as independent, aquatic animals. Conversely, if S. maculosa is exposed to alpine conditions, oviposition does not take place till the embryos have passed beyond the aquatic, gill-bearing phase. Therefore, in these circumstances, they are born as land-animals.

Hence the adjustment of each species to its environment is due to the direct effect of certain of the physical conditions of that environment on the course of development of the embryos. The fact of adaptation is not denied, but the mechanism whereby it is effected is discovered, and the way made clear for a fuller physiological analysis of the mode of reaction of protoplasm to physical stimuli.

The problem with respect to periodicity of egg-laying by Convoluta requires us to ascertain whether it is possible to refer the periodicity to any definite, recurrent physical condition of its natural environment.

The facts about to be related appear to indicate that this is possible.

It may be premised that if adult C. roscoffensis are kept in darkness for some time previous to the full development of their eggs, no egg-capsules are laid. The lack of egg-production on the part of dark-kept animals is due to the fact that animals kept under such conditions become starved and, as a consequence, incapable of supplying the eggs with food-materials. But if a similar experiment is made with animals containing eggs in an advanced stage of development and already supplied with plenty of food-materials, it is found that the number of egg-capsules produced by animals kept in darkness is actually greater than that produced by animals which are exposed throughout the day to the light. Hence we may infer that exposure to long spells of twelve or more hours of light is unfavourable to the maturing or deposition of eggs. Further experiments on similar lines show that egg-laying reaches its maximum when the animals are subjected daily to one short spell of six hours' light-exposure followed by a long spell of eighteen hours' dark-exposure. But—and the fact is remarkable—these conditions of light and darkness are precisely those to which C. roscoffensis is exposed during the spring tidal periods at which its eggs are laid habitually. At such periods, low water of successive tides occurs about the middle of the day and of the night, and hence, in twenty-four hours, the C. roscoffensis zone is uncovered once during day-time and once during night-time. So it comes about that, during the spring tides, C. roscoffensis is exposed for about six hours to the light and for the rest of the twenty-four hours is in darkness. Therefore, as the laboratory experiments show, of all the daily changing light conditions to which it is subjected throughout a lunar period, those which obtain at spring tide are most favourable to the deposition of egg-capsules.

In ascribing to light a leading *rôle* in determining the periodicity of egg-laying we have the support of not a few well-established biological facts. Thus the profound influence which light exerts on plants, both on their development in general and on their flower-production in particular has long been recognised. Perhaps the best-known example of this influence is afforded by the common ivy. It is a fact of general observation that ivy growing on a wall rarely if ever flowers, though when climbing over an arch exposed on all sides to the light it blooms freely. These effects of illumination on flower-formation have been investigated by Vöchting, whose researches are summarised by Goebel (1900). In order that plants may form flowers in a normal way, the illumination must not sink below a certain amount which is very unequal in different species. If illumination is allowed to sink below the required amount, the size of the whole flower or of its individual parts is diminished and, with decreasing illumination, a stage is reached at which the formation of flowers ceases.

Similar phenomena are doubtless common among animals though they have not been investigated systematically. Thus, though the phenomenon is not one of reproduction in the strict sense, we may cite Loeb's account (*loc. cit.*) of the effect of light in inducing regeneration of the polyps of the Hydroid Eudendrium racemosum. If a stem of this Hydroid, covered with polyps, is put into an aquarium, the polyps fall off very soon. If the aquarium is in darkness, no regeneration of the polyps takes place even after several weeks; but, when they are exposed to the light, new polyps form in the course of several days. We may suppose that light favours the formation of definite substances which are the pre-requisites for polyp formation.

Similarly, we are bound to conclude from our

experiments on C. roscoffensis that a spell of illumination of brief duration favours one or other of
the series of processes which results in egg-laying,
that a longer or shorter spell of illumination is unfavourable to this process, and that, when animals are
subjected to these unfavourable conditions, many of
them, though they are carrying eggs in an advanced
stage of development, remain sterile.

With respect to the periodicity of egg-laying by
C. paradoxa it is not so easy to refer the periodic
character of this event to the influence of light. It
is noteworthy that other littoral, marine organisms,
(certain brown algæ) living in almost identical habitats,
exhibit an identical periodicity with respect to their
reproductive processes.

This, according to Williams (1898), is the case
with the brown sea-weed Dictyota dichotoma, and
subsequent observers have shown it to be true of
other marine algæ. Not only does Dictyota liberate
successive crops of fertile eggs at fortnightly periods
but it sheds them at the same point in the tidal
period as that chosen by C. paradoxa for the discharge of its egg-capsules. In either case, the eggs
are liberated some three to five tides after the
greatest springs. During the subsequent tides of
smaller amplitude, the zone which forms the habitat
of both sea-weed and plant-animal is continuously submerged. Hence we can scarcely escape the conclusion

that the period selected for egg-laying has reference to the greater security which is offered during the first days of larval life. Born into the world at this period, the animals and the plantlets have some days of submerged grace before they become subjected twice daily to such extreme environmental changes as occur at other phases of the tidal sequence.

Thus, driven back provisionally on a teleological explanation, we may interpret the significance and origin of periodicity of egg-laying in the following way. One condition for survival of the species C. paradoxa and of Dictyota is that the just-liberated young shall be for some days after birth continually submerged: that, for one reason or another ultimately connected with nutrition, the maturing of these marine organisms and the development of their sexual cells requires a period of fourteen days, and that the organisms fit their fortnightly periods into the tidal periods in such a way that they reach their climaxes at the most convenient moments. As the waving flag of the guard gives the signal for the train to start, so change of light intensity appears to give the signal for the maturing of the sexual organs and thus secures their liberation at the proper moment.

Whether such bi-lunar periods of fertility exhibited by littoral, marine organisms have any bearing on similar periodic phenomena exhibited by the higher land-animals it is impossible to say ; though it is tempting to think with "The Lady from the Sea," "that we all are descended from sea-animals, and that if we had only accustomed ourselves to live our lives in the sea we should by this time have been far more perfect than we are."

CHAPTER II

THE fact which stands out most prominently from
open-air observations of C. roscoffensis and C. paradoxa is that the behaviour of these animals is
complex and purposeful. By some means or other
they create for themselves an ordered life, in spite of
the welter of change in their environment. Through
the ever-varying conditions of the world in which
they live, they thread their consistent way as surely
as we, with conscious self-control and agility, pick our
ways safely through the crowded traffic of the street.

We have now to endeavour to ascertain the
nervous components of the complex behaviour of
our plant-animals; to learn, by the method of experimental analysis, whether it is possible to refer the
ordered complexity of this behaviour to some few,
simple, nervous acts.

It is a matter of common knowledge that many
organisms, both plants and animals, orientate them-

selves with reference to the directions in which light and gravity act upon them. A geranium in a cottage window so disposes its leaves that they receive the maximum of such light as may reach them. Each leaf places itself at right angles to the direction of the incident light. The stem of the plant behaves differently, bending till its tip is parallel with the rays, it grows toward the source of light. Light is the agent, or stimulus, which induces these orientations. The mode of orientation is determined by the plant itself and has, in each case, a purposeful significance. The leaves in the window are none too well illuminated ; the work which they have to perform depends on ample light and thus, by their orientations, they secure for themselves the most favoured light-treatment.

The root of a plant grows vertically downward through the soil. When, from one cause or another, the tip is displaced from the vertical line, the rate of elongation of the growing region of the root becomes faster on the upper than on the lower side. A growth-curvature results, and the tip is carried by the bending root once more into the vertical line. Here gravity is the stimulus, and the result of the stimulus is a motor response—a purposeful growth-curvature. Cut away the root-tip, and the root, although it be displaced from the vertical, grows indifferently in any direction till a new root-tip is regenerated. From such and similar observations Charles and Francis Darwin (1880) concluded that the power of gravi-perception is localised in the root-tip. Nor has subsequent criticism succeeded in invalidating this conclusion.

Now, since the growth-curvature, which results from the perception by the root-tip of the stimulus of gravity, occurs in a region of the root—the elongating region—which is separated from the tip by a region which is not increasing in length, it follows that perception by the root-tip results in an excitation of the living substance of its tissues, and that this excitation gives rise to some change in the tissues which intervene between the perceptive and motor regions. This change, of unknown nature, we may call a nervous impulse, and we may say that, as the result of excitation consequent to perception, a nervous impulse is transmitted from the root-tip to the motor region. Of the nature of this impulse we know nothing; nor, for the matter of that, is anything definite known of the nature of any nervous impulse; for example, that which travels along a nerve to a muscle in one of the higher animals. The "nervous impulse" may well be of chemical nature, and transmission of such an impulse through living tissues does not connote definite specialised nerves. It is as much the property of protoplasm to transmit nervous impulses as it is of fire to burn, or of a lit fuse to explode a charge of gunpowder. Protoplasm is an apparatus for that purpose : as well, of course, as for other purposes.

Arrived at the motor region, the nervous impulse sets up an excitation in the protoplasm of that region. As the result of this excitation, there arises a definite modification in the hitherto uniform rate of elongation of the cells of the motor region. The cells of one side grow faster than those of the other, and a growth-curvature results by which the tip is carried back to the vertical position. Such a mode of nervous action is called a reflex. In every case, in the simplest, unicellular organism and in the highest animals, reflex action involves perception, excitation, transmission to the motor region, excitation of, and motor (or other) response by, that region. All protoplasm, as we know it, contains the apparatus required for this series of events, and evolution, as we know it, has but resulted in the perfecting and complicating of these reflex arcs. We may take the reflex as the base or primal manifestation of all nervous activity. In the reflexes of root and stem and leaf, the stimuli—light, gravity, etc.—which induced them give rise to the assumption by the root or stem or leaf of a definite position with respect to the direction whence the stimulus proceeds. Such stimuli are therefore called directive stimuli. When a fixed plant or animal responds to a directive stimulus by a definite, purposeful curvature we describe the response as tropistic. If the plant or animal is not fixed but free, it responds by moving in a definite direction and the response is described as tactic. Inasmuch as the end of either reaction is the achievement of a definite orientation and inasmuch also as the fixed plant, not only curves till it assumes a definite position, but also, having done so, moves by growth in the direction to which its curvature has brought it, we may use the term tropistic to describe the reactions of both fixed and free organisms to directive stimuli.

We have now to consider the tropisms of our plant-animals.

Brought into the laboratory and placed in sea-water in a glass vessel near the window, C. roscoffensis behaves precisely like the leaf of the geranium in the cottage window. Each animal turns to the light, moves toward it and finally exposes the surface of its body athwart the line of light. Within a minute or two the reaction is completed. Swiftly and, as it would seem, inevitably the animals assemble on the side of the vessel toward the light, and form a green scum on the surface of the water. If the vessel is turned round, the animals release their holds and, either falling like a precipitate to the bottom or edging round the side of the vessel, arrive once again at the water's edge on the side of the vessel directed toward the light. This mode of response we speak of as a positively (+) phototropic response.

C. paradoxa, which lives in a shadier situation, responds to the light of the laboratory by an opposite movement—a negatively (−) phototropic response (Fig. 9, *a*).

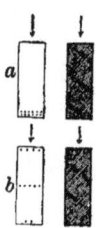

Fig. 9. Phototropism of C. paradoxa : the influence of light-intensity on phototropic response. *a*. Mode of response when the light-intensity is high. *b*. Mode of response when the light-intensity is low. The glass troughs containing the animals are represented (in plan) by oblongs. The troughs standing on a black ground are represented by the shaded, those on a white ground by the clear oblongs. The animals are indicated by dots, and the arrows show the direction of the light.

It is easy to prove, however, that neither the positive phototropism of C. roscoffensis, nor the opposite mode of reaction of C. paradoxa, is, in reality, an inevitable reaction. Expose C. roscoffensis suddenly to a bright light and it recoils (see Fig. 12)—as we ourselves in similar circumstances may recoil. Place it in a dim light and it exhibits no phototropistic response: it has become non-phototropic. On the other hand, in the gloom of an ill-lit cellar, in which the light-intensity approximates to that of its habitat among the masses of brown sea-weed, C. paradoxa becomes somewhat + phototropic (Fig. 9, *b*).

It is urged not infrequently that reflexes are the nervous units, unalterable in form, of which behaviour and higher phases of nervous activity are composed. The briefest study of the lower animals demonstrates that, though reflexes may be regarded as units by the physiological summation of which behaviour and habit are composed, they may not be regarded as unalterable and inevitable. No more than conscious acts are reflexes the masters of the organism which exhibits them. They are but servants, and tropistic reflexes serve the master-organism, to draw it this way or that according as it is well that, this or that route be taken. Under unchanging conditions, both with respect to environment and with respect to the state of the organism, the reflex is inevitable. But under such conditions a conscious act is likewise inevitable. Since unchanging internal and external conditions are all but unknown in nature, there will always be scope for modification in the reflex as in the conscious act. If the physiologist is called in to act as umpire in the dialectical game between the advocates of free-will and determinism, he pronounces the game a draw. Under abnormal and well-nigh impossible conditions, the organism, high or low, is an automaton—the creature of inevitable nervous responses—reflex or conscious. Under normal conditions of life, it responds now this way and now that to external or internal stimuli and so appears to act as a free agent.

The apparent inevitability of reflexes is but an indication of habit. When the environmental circumstances to which an organism is exposed are comparatively simple or when the organism itself is not highly differentiated, one or two external agents may serve it as guides. The organism takes the habit, for example, of relying implicitly on the stimuli of light and gravity. By responding to these stimuli, it finds its proper place with such certainty that other modes of response to other stimuli are ignored habitually. Hence, by playing on its habitual tropisms, it is easy in the laboratory to lure an organism to its doom.

This we may illustrate by exposing C. roscoffensis to simultaneous stimulation by light and heat. It must be premised that the animal, though, for a marine organism very tolerant of high temperatures, is negatively thermotropic at about 35° C. At this temperature it moves in the direction of the colder water. In order to investigate the behaviour of the animal with respect to simultaneously applied light-

and-heat stimuli, large numbers of C. roscoffensis are placed in sea-water contained in a long glass trough, the axis of which is parallel with the direction of the light. Promptly, the animals mass themselves at the end of the vessel directed to the light. The water at that end is heated gradually ; but in spite of the rising temperature, and in spite of its powers of negative thermotropism, Convoluta remains faithfully at its light station, and dies in thousands at its post. Habitual obedience to the command of light renders it oblivious of the warning of increasing temperature, which warning suffices to bring about the withdrawal of less pre-occupied animals from dangerous regions. We see in the behaviour of the plant-animals thus subjected to simultaneous stimulation not an illustration of the inevitableness of a reflex, but an example of the limitations attaching to all nervous actions, both reflex and conscious.

The behaviour of C. roscoffensis with reference to black and white backgrounds supplies a striking illustration of the fact that circumstances alter reflexes. These "background" responses we will now consider. The choice of a definite background is a phenomenon exhibited by many sea-shore and aquatic animals. When offered the alternative of a white or black background, some animals take up a position on the one, some on the other. Thus among the marine Crustacea, certain prawns and also species of

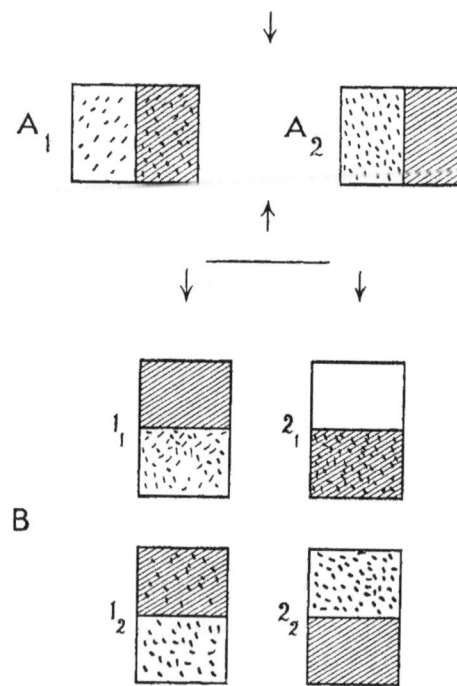

Fig. 10. Diagram illustrating the influence of background (white or black) on the movements of Convoluta roscoffensis. The animals are placed in shallow porcelain troughs, the bottoms of which are half white and half black (shaded). Each dash represents a Convoluta. **A.** In uniform light. **A₁.** At beginning of experiment, Convoluta fairly uniformly distributed. **A₂.** After forty minutes, Convoluta all on white ground. **B.** In lateral light (arrows show direction of light). Fifty Convolutas placed in the white half of 1_1, and fifty in the black half of 2_1. 1_2 and 2_2 show the results after two minutes :—in 1_2 ratio on black and white $= \frac{20}{30}$; in 2_2 ratio $= \frac{0}{50}$. See text.

Mysis station themselves on the black part of a dish the bottom of which is half black, half white. The chameleon shrimp, Hippolyte varians, on the contrary, selects the white ground (Fig. 10). A similar behaviour is exhibited by various fishes, trout among others.

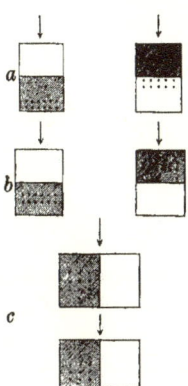

Fig. 11. Phototropism of C. paradoxa: the influence of background on phototropic response. The flat, porcelain troughs containing the animals are represented (in plan) by the oblongs. The bottom of each trough is half white and half black. In the diagram, the white ground is indicated by the unshaded, the black ground by the shaded part of the oblong. The animals are represented by dots, and arrows show the direction of the light. *a.* In bright light. *b.* In weak light. *c.* "Choice" of black ground in preference to white ground.

So striking is the behaviour with respect to background both of C. roscoffensis and C. paradoxa,

that when specimens of the two animals are put
together into a dish, the bottom of which is half
black and half white, they segregate rapidly and
completely ; C. roscoffensis takes up positions on the
white ground, C. paradoxa on the black ground
(cf. Figs. 10 and 11). The distribution is in accord
with reasonable expectation based on knowledge of
the natural habitats of the two species. C. roscoffensis, attuned to a high light intensity, with its
place in the sun, is evidently a bright background
animal ; C. paradoxa, lurking in the shadows of the
weeds, though it also needs light for its growth and
development, is unused to well-lit situations and
seeks in preference the darker background.

But though the selection of ground seems biologically reasonable the question remains, how is it
done ? The hypothesis may be hazarded that it is
a phenomenon of association. C. roscoffensis is, as
we know, positively phototropic. In effecting a
phototropic response, it is bound in ordinary circumstances to pass from a darker to a lighter
background. The performance of the phototropic
movement is associated with the darker ground, the
achievement or consummation—that is, a state of
immobility—is associated with the brighter background. If therefore we adopt the hypothesis,
proposed by Semon, (1904) that environmental conditions, which are contemporaneous with a particular

stimulus, are recorded in the "mneme" or unconscious memory of an organism as integral parts of the nervous operation initiated by the stimulus and consummated by the reaction which it calls forth; then it may well follow, as it follows in organisms endowed with conscious memory, that these environmental conditions acting alone and in the absence of the stimulus, may suffice to set in action the nervous apparatus in the same manner as the stimulus itself originally set that apparatus in action. Hence the attendant environmental conditions may produce the reaction originally called forth by the stimulus.

An example borrowed from Semon's work (*loc. cit.*) may make the idea clearer. A boy throws a stone at a puppy. The dog is hit and hurt, whimpers and runs away. The next time the puppy—grown older and wiser—sees a boy stoop, as though to pick up a stone, it whimpers and runs away. Linked in the memory are the hurt, the stone, and the stooping boy. The hurt supplied the stimulus for whimpering and flight; but memory, the constable of the body, charges the stooping boy with being an accessory to the act. Henceforth it will advise the avoidance of stooping boys. Experience consists in the discovery of short cuts to safety.

So, assuming with Samuel Butler and Hering, an unconscious memory, or mneme, Semon suggests that the lower organisms may react to the attendant or accessary stimulus in the absence of the principal.

Applying this hypothesis to background reaction, we assume that dark background, from constant association with unilateral light, has come to suffice to stir up the mneme and so to set going the nervous apparatus which induces a precise muscular movement. Thus, C. roscoffensis, placed on a dark background, begins to crawl about and continues to do so. That this interpretation of the origin of background reaction, contains something of the truth seems probable from the fact that the movements performed by the animals on a dark background are, compared with the business-like, phototropic movements, aimless as to direction. They are non-directive, chance movements ; but since they continue so long as the dark background is there to call them forth, they conduct the animals sooner or later to the white ground of the particoloured vessel. Arrived there, the stimulus ceases from troubling and C. roscoffensis is at rest.

We conclude, therefore, that background, from being a mere attendant circumstance, an environmental accessory to unilateral light, has come itself to serve as a stimulus to movements which, by the devious paths of chance, direct the animals to the lighter ground.

In nature, under all ordinary conditions, back-

ground co-operates with unilateral light to bring
C. roscoffensis to its proper light station—its place
in the sun. Nevertheless background stimulus may,
under artificial conditions, act antagonistically to
that of lateral light and even dominate it.

Thus if two half-black, half-white porcelain
troughs containing a little sea-water are so placed
that, in one, the white half, in the other, the black
half is directed towards the source of light, then, if
some fifty C. roscoffensis are placed in each of the
troughs, whereas in a very short time all the animals
are congregated on the white ground of the first
trough, only about forty per cent. manage to arrive
at and maintain themselves upon the black half of
the second trough (Fig. 10, *B*).

When positive phototropism involves the passage
from black to white the movement is executed with
certainty and despatch ; but when it demands a
passage from white to black—a movement against
the grain of habit—there are hesitation, uncertainty,
and many failures. Under yet other conditions,—
when, for example, the intensity of the unilateral
light is lowered—the stimulus of background may
prove more potent altogether than that of unilateral
light. In such circumstances, C. roscoffensis remains
on the black half of the dish, although the rays
of light signal to it to approach their source. It
looks as though the facts or illusions which, in

higher animals, are named choice and volition are illustrated in our plant-animals by the simplest of working models. If this view is accepted, it would seem to follow that the intricacy and mystery of complex habits and instincts are begotten of the ever-increasing complexity of conditioning or accessory stimuli which have first attached themselves to and then replaced the original series of stimuli.

Some further facts with respect to the phototropic responses of C. roscoffensis are worthy of a passing word.

In the first place, just-hatched animals, though they respond to the directive stimulus of gravity, do not respond to that of light. After a few hours of free existence, they acquire the faculty of responding tropistically to unilateral light, which henceforth becomes a masterful factor in determining their habits.

In the second place, the rays of light to which C. roscoffensis responds tropistically are not those which induce phototropic curvatures in plants. As is well known, a plant exposed, unilaterally, to rays of the less refrangible part of the spectrum—the red for example—shows no phototropism; whereas a plant subjected on one side to blue-violet light reacts as readily as to white light. Convoluta roscoffensis, on the other hand, responds not to blue or violet light, but to green light. The diagram (Fig. 12) represents

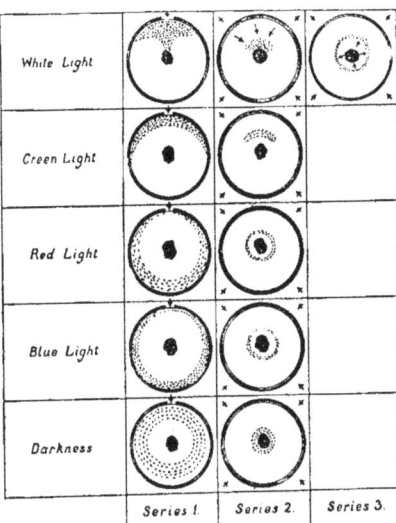

Fig. 12. Tropistic reaction of Convoluta roscoffensis to Monochromatic Light. Each circle represents a ground-plan of a shallow
porcelain vessel containing a central heap of sand, a little sea-
water, and many Convolutas (represented by dots). The break
in the circle (Series 1) indicates the position of the window in
a blackened bell-jar placed over each porcelain vessel. The
arrows represent the direction of the light. Series 1 shows
the disposition of Convoluta at the time of removal of the
covers. Series 2 shows the instant negative phototropic reaction
set up by removing the covers (raising the light intensity).
Series 3 (only one example shown) shows the recovery (a few
seconds after Series 2) of the positive phototropism. The green
light was produced by passing daylight through three of Baker's
green gelatine films ; the red by using three of Baker's red films ;
the blue by using four of Kirchmann's blue films and one
green film. (The blue and green lights were not absolutely
monochromatic.)

the results of a series of experiments with unilateral, monochromatic light. As the illustration shows, the animals—represented by dots—mass themselves toward the source of light when that light is white or green. In blue light, they remain distributed with fair uniformity around the periphery of the containing vessel. In red light they show some sign of a negative reaction.

The most probable explanation of this response to green light is that the orange eye-spots and pigment glands are the organs of light-perception; and that the pigment contained in the eye-spots and glands absorbs principally the green light. From observations on other animals it would appear probable that green light not infrequently serves marine organisms for purposes of perception: nor, when we reflect upon the green-blue colour of sea-water, will this appear surprising.

In the third place, C. roscoffensis, with its well-defined tropisms, is an admirable subject in which to study what without a great violation of language we may call the problem of the parallelogram of physiological forces: in other words, the problem of the mode of response of an organism to two directive stimuli, simultaneously applied and acting along different lines. As might be expected, when two stimuli act on C. roscoffensis one not infrequently dominates the other, so that the resulting reaction is that which would occur were the dominating stimulus alone applied. This happens, as we have seen, when C. roscoffensis is subjected to both light- and heat-stimulation.

So also, in the case of light and gravitational stimuli acting simultaneously, the mode of response of C. roscoffensis shows that the latter stimulus may be ignored.

Thus, if animals are placed in water in a tall glass cylinder on a steady table, they rise to the surface of the water and congregate on the side toward the light. If the light-conditions are modified by the interposition of a black card or plate of ground glass between the source of light and the top of the water, C. roscoffensis relaxes its hold and swims downward to take up a new position just below the edge of the screen. Geotropism is subordinated to phototropism.

Subjected to simultaneous stimulation by light and gravity, C. roscoffensis behaves exactly like a green plant placed under similar conditions. Though the stem of a green plant is negatively geotropic, yet, if it is illuminated from below, the plant, ignoring gravitational stimulation, directs the tip of its stem downwards toward the source of light.

The behaviour of both plant and animal would seem to indicate that, in the reflex groundwork of nervous activity, something akin to the phenomenon of attention in psychic life may manifest itself.

Even though he may not be concerned with problems of reflex action, the biologist who would investigate the life histories or structure of such animals as C. roscoffensis must pay some heed to their tropistic behaviour, for on this knowledge his successful manipulation of the living animals depends.

Thus, to obtain plentiful supplies of eggs we may make effective use of the phototropic reaction. Large numbers of mature C. roscoffensis are placed near a window in a flat, glass dish containing sea-water. The animals move up to the light. At nightfall, the dish is turned round. The operation, if performed carefully, does not disturb the animals. They remain throughout the night in that part of the vessel which was nearest the light during the previous day. Certain of the animals deposit egg-capsules. In the morning, the animals, responding to the directive stimulus of light, cross over to the side turned toward the window. The egg-capsules thus left behind are readily visible to the eye and may be picked out by means of a pipette. In this way, several hundred egg-capsules may be obtained in the course of a few days.

Again, young C. roscoffensis are so minute that they may be found only by a practised eye. Nevertheless, by exploiting their background reaction, they may be picked out easily.

By laying a small sheet of white paper on a black cloth and standing the dish containing the animals partly on the black and partly on the white ground, the animals are caused to accumulate above the latter. There, however, they are almost invisible; but by turning the vessel round so that the part above the white paper is brought over the black cloth, the animals may be seen distinctly and picked out by means of a fine pipette before they scuttle off again to the white ground. To transfer a young C. roscoffensis from one vessel to another is difficult enough till its geotropism is pressed into service, when the operation becomes quite easy. The animal is lifted in a pipette; but on endeavouring to expel it from the tube by pressing the indiarubber nipple of the pipette, only the water is discharged, and C. roscoffensis is left sticking to the side of the glass tube. To avoid this, the pipette is held vertically and no effort made to eject the water. The slight, involuntary shaking of the hand suffices to render the animal negatively geotropic. Down it swims till it reaches the drop of water at the point of the pipette, whence the gentlest pressure suffices to transfer both drop and animal to another vessel.

So far, this study of the behaviour of our plant-animals has been confined to the investigation of their tropistic responses to light and gravity. A stimulus may however induce responses in an organism of a very different kind. It may give rise, not to a change of place, but to a change of *state*. Such effects of stimulation are called tonic effects, and the organism which responds to them is said to be in a state of tone or tonus. Certain peculiar effects of this kind are well known among human beings and may serve us as illustrations. People who work all day by artificial light, specially by unmitigated electric or incandescent gas light become irritable and depressed. Professional photographers, who spend long hours in "dark rooms" developing photographs in red light, suffer mentally in a similar way. The change of state induced by such abnormal conditions we may describe as a change of tone. We will assume that light is indispensable to the human race, that men's bodies are attuned to light and that this harmony is maintained by an unceasing sequence of light-stimuli which contribute to the well-being of the nervous system. Then, if we adopt this view, it will be easy to imagine that a cessation of the rain of stimuli may prejudice the well-being of the nervous system and be the origin of disorders of more or less severity.

How far the normal nervous state of human beings is determined by the tonic effect of light it is not possible to say; but there is no doubt that

this phototonic effect is of considerable and even
fundamental importance to the well-being of many
animals and plants.

If a green plant is placed in darkness the
mechanism of its growth is thrown out of gear.
Though it grows, and, if supplied with proper food,
might, for all we know, go on living indefinitely, its
nervous state is changed, its tone has been affected.
It becomes "drawn," as gardeners say, its stem grows
long and supple, and its leaves remain small and un-
developed. As, without the controlling baton of the
conductor, the unity of the orchestra is lost, and, it
may be, harmony is replaced by discord, so, without
the constant influence of light, the harmony of growth
which obtains normally throughout the plant is dis-
turbed. Since, therefore, light-stimuli contribute to
the maintenance of the normal nervous state of the
plant, we say that light exerts a tonic influence. Inasmuch, however, as even in the absence of light
the plant remains alive and capable of some sort
of growth and development, we must conclude that
the state of tone in which it lives is not the result
of light but that it is modified by light. This
effect of light in modifying—to the advantage of the
organism—its state of tone is called phototonus. The
language is clumsy but the ideas which it conveys
are clear enough, though lacking in precision.

C. roscoffensis is, as we know, attuned to a high

light intensity. It exposes itself on the beach to the bright light of the midday sun with nothing between it and desiccation but a constant, filmy stream of salt, drainage-water. In such situations, it retains its powers of activity unimpaired for hours. If, however, it is placed in a vessel with some sand and water, taken to the laboratory and kept in darkness, it passes after some days into a lethargic condition. In this state of dark-rigor C. roscoffensis remains on the surface of the sand and may even fail to respond by downward migration when subjected to the stimulus of vibration.

So also, after prolonged exposure to high light intensity, a similar lethargic condition—a light-rigor—comes over the plant-animals. Even in their natural positions on the beach, after long hours of exposure to the sun's glare, colonies of C. roscoffensis may be observed in which all the members appear to be overcome by light-rigor. They lie roped together by the slimy excretion of their skin, inert, floating in water-puddles. At times, chunks of a colony in this state may be detached by running water, and small green masses, each of many thousand individuals, are borne seaward by the drainage stream. In this lethargic state of light-rigor, which both young and old animals exhibit, C. roscoffensis is difficult to manipulate; for example, attempts to transfer them from one vessel to another by means of a pipette result generally in damage to the animals. To this

lethargy or light-rigor—here attributed to exposure
for long periods to high light intensities—is due the
fact that, whereas, on some days, the C. roscoffensis
patches on the shore disappear before the water of
the making tide reaches them, yet, on other days,
the multitude of animals composing a patch lies
motionless and indifferent to the approach of the incoming tide.   Not till the first wave sweeps over them,
do the animals throw off their sloth and disappear.

We have now to attempt to apply the knowledge
we have obtained of the tropistic responses of C.
roscoffensis to light and gravity, and of the tonic
effects of light, to the elucidation of the most strikingly
picturesque feature of the behaviour of this animal,
that of its tidal uprising and downlying.   Almost as
soon as the water of the falling tide has run off the
roscoffensis zone, the green colonies appear, and,
before the making tide invades it, they vanish.   The
purposes of ascent and descent are obvious.   By its
ascent, the animal reaches the light without which—
for reasons we shall discover subsequently—it cannot
live ; by its descent, C. roscoffensis maintains its
situation on the shore and escapes the waves.

As our study of its tropisms makes clear, these
movements of ascent and descent may be induced in
the laboratory by subjecting the animals to appropriate stimulation.   Vibrations produced by tapping

the sand or the containing vessels send them down,
only to reappear when the tapping has ceased. But,
as we have seen, a colony may disappear before the
tide has mounted high enough to disturb it. The
eyes of C. roscoffensis—mere pigment spots—are too
rudimentary to allow us to suppose that it sees the
water coming in and so takes warning and descends
betimes. Indeed a simple experiment suffices to
demonstrate, not only that this is not the case, but
also that whole colonies of C. roscoffensis may descend
beneath the sand in the total absence of an apparent
external stimulus.

Thus, if a batch of animals from a roscoffensis patch
is scooped up with sand and water by means of a cup
and taken into the laboratory, the shaking to which, of
necessity, the specimens are subjected in the process
causes their swift descent. By the time the cup is
brought indoors, not a trace of green may be visible
in it ; but, in the calm of the laboratory, the animals
reascend once more and lie as a thick, dark green
scum upon the surface of the sand. They remain in
this state for hours, then suddenly disappear.

Wondering at this swift retreat, and as we wonder,
staring through the laboratory window at the shore
a stone's throw away, we note first vaguely and then
with quickening curiosity that the rising tide is just
about to flood the roscoffensis zone. Curious, this
descent of the green scum of animals in the cup !

Some hours elapse, and, as the tide is running off
the roscoffensis zone, curiosity, or its after-effect,
provokes us to inspect the cup on the laboratory
table. Even as we look into it, a faint green colour
steals over the surface of the sand, and, in a minute
or two, it is almost black with a dark crowd of
C. roscoffensis. Now curiosity joins with astonish-
ment to beget a new idea. More cups are found
that the observation may be repeated and coincidence
put out of court. Each time we repeat the observation
on fresh batches of animals we obtain the same result.
As on the shore in the roscoffensis zone, so in the
laboratory the upward and downward movements of
Convoluta march with the movements of the tide. As
the tide recedes from their home upon the shore, the
sojourners in the laboratory rise up : as the tide rises
over it, they sink down. In the absence of all apparent
external stimulus, C. roscoffensis, obedient to its
custom, yet keeps time with the tide. The rhythm of
the tides is reflected by the movements of the animal.
For eight successive tides (Fig. 13) the animals in the
laboratory maintain their rhythm, synchronous with
the ebb and flow of the waters over the roscoffensis
zone : then, though the rhythmic movement up and
down may yet continue, its temporal periodicity
loses precision, and, finally, the rhythm is worn down.
This stage reached, the animals exchange a working
day of double, six-hour shifts, two up, two down, for

one of a single, twelve-hour spell of "upness" with a like twelve-hour spell of "downness." In other words they phototrope themselves up to the light as day breaks and sink down with the sun.

Whence comes the power whereby C. roscoffensis acts as a tide-indicator? What orders its rhythmic coming and going?

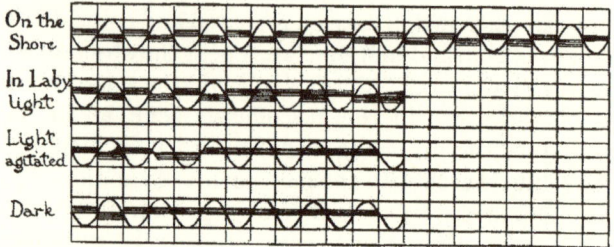

Fig. 13. The rhythmic tidal movements of C. roscoffensis. The curves represent the rise and fall of the tide. The horizontal lines included within the tidal curve indicate the "up" or "down" positions assumed by the animals. "In Laby. light" shows that, with animals kept in the laboratory and exposed to light during the day, the rhythm is lost after seven or eight periodic tidal movements up or down. "Light agitated" shows that animals exposed to constant vibration lose their periodicity more quickly. "Dark" that in constant darkness periodicity does not manifest itself.

A French biologist, Dr Bohn (1903), who has also observed this periodicity of upward and downward movement, rejects the view which is put forward

below, and regards the phenomenon as a manifestation
on the part of C. roscoffensis of "memory of the shock
of the waves." Certainly, if all other explanations fail
—if we can discover no agent which serves to jog this
memory—we must accept this suggestion; though in
doing so, it might be well to ask ourselves whether it
is to be regarded as an explanation or as a succinct
statement of our ignorance.

Experimental investigation of the phenomenon
would appear to indicate that no such large demand
on memory—or mneme—as that which is implicit in
the above hypothesis need be made.

In the first place, as we have noted already,
C. roscoffensis does not remain on the surface of
the sand at night. Hence we must suppose, on the
memory hypothesis, either that it forgets to arise from
the dark sand when it is dark on the surface, or that
it remembers, rises, and finding nothing better to do,
goes to bed again.

The behaviour of C. roscoffensis in constant darkness is yet more difficult of interpretation on this
hypothesis. For, when kept in continuous darkness, C. roscoffensis ceases to exhibit periodicity of
alternate up and down movement. There may be
one movement downward and one upward; but, after
that, the animals remain upon the surface of the
sand day after day until they die (Fig. 13).

Again, if the downward movement is due to a

K.   5

memory of past vibrations caused by the making tides invading periodically the C. roscoffensis zone, how much more certain should be the effects of present vibrations. Yet, if the vessel containing the animals is so exposed that a steady drip of water falls upon the surface of the sand contained in the vessel, C. roscoffensis clings to its periodic habit. As soon as it perceives the vibrations it descends and remains below the sand. When, however, the time for its uprising arrives, it rises to the surface, and, in spite of injuries, remains upon the surface. It seems difficult of belief that the memory of a particular kind of blow can be a more powerful spur to action than the actual receipt of an unceasing series of blows of a like kind. The original suggestion which, though it is not accepted by the author of the memory hypothesis, seems to fit the facts, seeks to explain the periodicity of upward and downward movement exhibited by C. roscoffensis by connecting it with tonic light effect.

In support of this it may be mentioned that C. roscoffensis fails to exhibit its tidal rhythm except when it is subjected to a fairly high light intensity during its period of "upness." Thus, even in a room at some little distance from the window, the movement does not keep tidal time.

Again, other observations indicate that the spell of illumination counts for something in determining the precision of the movements. Thus, if three batches of C. roscoffensis, collected directly after the colonies emerge, are put in darkness for periods of one, two, and three hours respectively, and are then exposed to the light, that which had only one hour's run in darkness descends first, and that which had two hours' darkness descends next.

Taking the results of these experiments into consideration and bearing in mind the condition of lethargy which C. roscoffensis may manifest, in its natural station, after long light-exposures, we are led to frame some such hypothesis as the following, in order to account for the periodic tidal movements exhibited by this animal.

Phototropism and background reaction lead C. roscoffensis to the most illuminated parts of that region of the beach which provides it with a continuous, filmy stream of water.

Independently of its tropistic effect, light exerts a tonic effect on the physiological state of the animals. Under the combined influences of tropistic and tonic light-stimuli, C. roscoffensis is held—at attention—in the "up" position: in other words, whilst subject to this constant rain of phototonic stimuli, it remains negatively geotropic. True, if the sand is agitated, the vibrations set up suffice to change the sign of its response to gravity and send it geotroping. Nevertheless it is easy to show that the response of C. roscoffensis to the vibration-stimulus is less marked at the beginning of an "up" phase than it is toward the end of that phase. Thus, if specimens are collected as soon as the tide is off the colonies and are brought in a vessel into the laboratory, they swarm up to the surface almost as soon as the vessel ceases to be shaken, whereas animals collected after a long light-exposure and placed in a similar position, may remain down till the next tidal "up" phase is due.

Thus it is reasonable to conclude that, after some five or six hours of light-stimulation, internal changes are induced which act as stimuli and cause the animal to change the sign of its response to gravity. It becomes positively geotropic and descends beneath the sand. In the darkness of the sand, recovery of the original, normal state takes place gradually, and the animals now respond to the stimulus of gravity by a movement in the opposite sense. They ascend to the surface. In its simplest form, the hypothesis involves the assumption that prolonged light-exposure and prolonged dark-exposure modify the tone or state of nervous irritability of the animals, and that these changed conditions manifest themselves by a changed mode of response to gravitational stimulus. After a prolonged light-exposure, the animals are positively geotropic; after a corresponding sojourn in the dark, they become negatively geotropic. The reversal of the direction of a tropistic movement is by no means unusual among plants and animals. Thus, in order to cause horizontally growing lateral roots to take up vertical positions, it suffices merely to remove the main root. As a result of the operation, the physiological state of the whole root-system is so changed that members formerly transversely geotropic become positively geotropic, and tertiary roots which previous to the operation were ageotropic (non-geotropic) and hence grew indifferently in any direction, become transversely geotropic.

Similar changes in sign of tropistic response may be induced by definite changes in the environment. For example, as Loeb has pointed out, freshwater Copepods, (small Crustacea) taken from the same pond at the same time, may exhibit, some a positive, some a negative, phototropic response and others may be non-phototropic. If, however, a little carbon-dioxide is added to the water they all become positively phototropic. It is not improbable that this uniform migration of the animals in the direction of the light which follows on the addition of carbon-dioxide is an example of response to associated stimuli. Copepods feed no doubt on algæ, which can only live and grow in the light. In the course of their nutrition, algæ decompose carbon-dioxide and liberate oxygen, so that the amount of carbon-dioxide contained in the water in their immediate neighbourhood is less than that contained in the darker regions of the pond. Much carbon-dioxide will be correlated with limited food supply. Now it has been shown definitely in the case of other animals, e.g. the caterpillars of Porthesia, that they are only positively phototropic so long as they are not fed. If this holds good for Copepods, their response to increased carbon-dioxide becomes at once intelligible on the mneme or associated stimulus hypothesis. Thus hunger affects the tone or physiological state in such a way that the Copepods respond to light by directive movements whereby food supplies become available. The movement brings the animals from a part of the water which contains a maximal amount of carbon-dioxide to a part where, thanks to the presence and activity of the green algæ,—the food sought by the Copepods—the water is not fully saturated with carbon-dioxide. When the animal encounters carbon-dioxide conditions which are normally associated with hunger conditions, it takes the hint and phototropes just as though it were hungry. For a hungry man, a cook-shop window has an irresistible attraction, whereas to the well-fed person it may offer no seduction, or even be repulsive : nevertheless, "si par impossible" the odour which emanates from it is very agreeable, the well-fed may deign to sniff.

What internal changes, chemical or other, resulting from the prolonged light-exposure of C. roscoffensis on the beach, give the signal for its dismissal from the attitude of attention which it takes up during the "up" period we do not know. Nor may we say with confidence that the explanation of the periodic rhythm which we have offered is complete or final. The subject deserves more detailed study than it has yet received, both for its own sake and for the light which it may throw on the origin of habit and, it may be, also, of instinct.

PART II

THE NATURE OF THE PLANT-ANIMALS

CHAPTER III

THE GREEN CELLS OF CONVOLUTA ROSCOFFENSIS
AND THE PART THEY PLAY IN THE ECONOMY
OF THE PLANT-ANIMAL.

IT is not only on account of their behaviour, as
exhibited by the tropistic movements and periodic
phenomena which we have recorded, that the plant-
animals C. roscoffensis and C. paradoxa attract the
attention of the biologist. The most superficial
microscopic examination is sufficient to convince
him that their tissues are not like those of other
animals. The green cells of C. roscoffensis and
the yellow-brown cells of C. paradoxa arrest his
attention. In regular and close rows, just beneath
the surface of the body, lie the green cells of
C. roscoffensis, each so minute as to be invisible to
the unaided eye and yet so numerous as to be the
source of the dark, spinach-green colour of the
animals (Frontispiece and Fig. 14). Though less
numerous and less regularly arranged, the yellow-
brown cells which lie beneath the skin of C. paradoxa
are, like the green cells of the former species, striking

and puzzling objects (Frontispiece and Fig. 15).
Puzzling because, whilst they seem to be just as
much integral parts of the bodies of the animals as
any other tissue-elements, they have nevertheless
a foreign and plant-like appearance. So plant-like

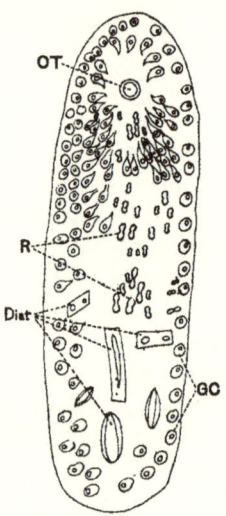

Fig. 14. A young Convoluta roscoffensis. GC = green cells. Diat.
and R = remains of diatoms ingested and digested. OT = otocyst.

indeed is the aspect of C. roscoffensis as seen under
the microscope, that a botanist might well be excused
for mistaking it for a fragment of a leaf, endowed
with an uncanny kind of movement.

In yet another and no less remarkable way,
C. roscoffensis exhibits a plant-like character. The
bodies of normal, mature animals never contain the
slightest trace of food-substances. Though it is
kept for days in pure sea-water till any ordinary
marine animal would be ravenous,—in point of
fact most marine animals are always ravenous—an
adult C. roscoffensis makes no attempt to ingest any

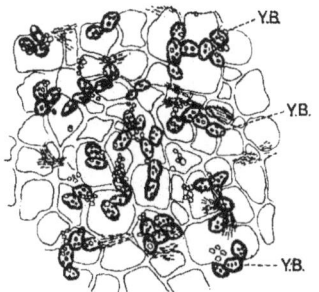

Fig. 15.   The superficial tissues of Convoluta paradoxa.
Y.B. = yellow-brown cells.

food-substances which may be added to the water.
Though tempted with diatoms, green algæ, starch
grains, oil drops, milk, or lamp-black, it remains with
its capacious mouth so pursed up as to be invisible
and refuses to ingest any solid food whatever. Till
last year it seemed that there was no exception to
this fasting habit of adult C. roscoffensis ; but during

observations on animals which had been kept for a month in darkness in pure sea-water, certain individuals were discovered which had so far condescended from this ascetic mode of life as to have become cannibals. Instead of being straight and slim, they carried a large pouch-like distension about the middle of their bodies. Microscopic examination showed that the pouch was occupied by another adult Convoluta as large as that which had engulfed it. Hence it follows that normal adult C. roscoffensis in its natural state does not refrain from food because it cannot swallow, but because it does not want to eat.

Now green plants do not take up solid food: they manufacture it. From inorganic substances, water and carbon-dioxide, which are absorbed from without, the green cells of plants manufacture sugar. This process, which is a preliminary to nutrition, is termed by botanists, photosynthesis, since the energy required for the manufacture of the carbohydrate (sugar) is derived from the radiant energy of light. The green pigment, chlorophyll, which is associated in the green cells of the plant with specialised, granular bodies called chloroplasts, absorbs light, and in some way, as yet imperfectly understood, this radiant energy is utilized by the protoplasm of the chloroplasts in the manufacture of sugar. The plant possesses also the power of synthesising yet more complex substances. Beside carbohydrates such as sugar, which consists of carbon, hydrogen and oxygen, the plant prepares synthetically its own nitrogenous food-substances, the proteins. Though next to nothing is known of the details of protein-synthesis as carried on by the plant, this much is known, that the nitrogen contained in the proteins is derived by the green plant from inorganic sources, chiefly from nitrates which are absorbed in solution from the soil or water in which the plant is growing. Having thus manufactured its food-substances from raw, inorganic materials, the plant is free to feed upon them, that is, to use them either to build up and repair its living substance (protoplasm) or to convert them directly or indirectly into substances (secretions) which enter into the composition of its tissues. Thus, for example, from the photosynthesised carbohydrate, are derived the cellulose substances which form the enclosing shell or cell-wall within which is contained each individual mass of protoplasm which we call a cell or protoplast. But beside serving such constructive purposes, much of the manufactured food-substance, particularly the carbohydrate material, is used for respiratory purposes, that is, for supplying the energy wherewith the plant does the work of living. By inducing compounds like sugar to unite with oxygen, their decomposition and oxidation are effected, with the result that energy is liberated and simpler substances, e.g. carbon-dioxide and water, are produced. The liberated energy serves for the performance of the work which the living plant must do, and also, converted into heat, contributes to maintain the temperature of the plant's tissues at a proper level. The surplus of carbohydrate and of protein not used for constructive or respiratory purposes the plant puts by for future use. The starch, oil and nitrogenous substances contained in seeds, tubers, and other storage-organs of plants represent this reserve food-material.

The power possessed by the green plant of manufacturing food-materials in excess of its immediate needs is the lever which makes the whole world of animal life to move. For the animal has no such synthetic powers, and yet it requires the same food-substances as the plant. Hence it is constrained to take them from the plant. The aphorism "all flesh is grass" is no mere figure of speech, but a terse statement of truth.

Though the foregoing facts are, of course, the commonplaces of plant-physiology, yet they require mention here, for it follows from them that, if C. roscoffensis does not take in solid food, it must either absorb it in solution or manufacture food for itself. Since the plant-animals not only live very well but also increase and multiply in pure sea-water, and since pure sea-water contains but the merest traces of any organic substances which might serve them as food, we are driven to accept the latter alternative, and to conceive of C. roscoffensis as an animal which lives like a plant, in other words, as a plant-animal. This conclusion forces us to direct our attention to the plant-like green cells which form such a prominent tissue in the body of C. roscoffensis.

From the general considerations which we have just advanced, it would appear to follow that the green cells possess the power, common to those of green plants, of manufacturing carbohydrate food-materials from the simple, inorganic, soluble substances, water and carbon-dioxide, and possibly also of manufacturing complex nitrogen-containing food-substances, such as proteins, from simpler bodies. If we succeed in proving that the green cells of C. roscoffensis possess these powers, other problems will present themselves. Thus, we shall want to know, what *are* the green cells? Is C. roscoffensis born with them or does it acquire them? If it acquires them, how do they get into the body and what are they like before they become constituents of the body of the animal?

As a preliminary to the investigation of these and other problems on the origin, significance and fate of the green cells, we will turn back to consider further the behaviour of C. roscoffensis and C. paradoxa with respect to food. The various observers who have occupied themselves with investigations into the mode of life of C. roscoffensis have all reached the conclusion that this animal does not take up solid food. A similar apparent total abstinence has been recorded in cases of other animals which contain green or yellow cells not unlike those which occur in our plant-animals. Thus no food has been seen in the bodies of certain adult Radiolaria, Ciliata, Hydrocorallines and Madreporaria, and in all these animals from which remains of food are absent, coloured cells are present. Hence the natural inference has been drawn that such animals subsist on the food-materials manufactured synthetically by their green or yellow cells.

If, however, the evidence which we have now to bring forward with respect to C. roscoffensis is applicable to the other green- or yellow-celled animals, then, though the conclusion may contain a large measure of truth, the premise on which that conclusion is based is erroneous.

When referring to the abstemious habit of C. roscoffensis we were careful to state that it is the mature animal which does not take up solid food. As a matter of fact, from the time of hatching to the period of maturity, C. roscoffensis feeds, and feeds voraciously. Indeed, its catholicity of taste is remarkable. Diatoms, unicellular algæ, spores of various kinds and, in the absence of more nutritious substances, grains of sand are swallowed with avidity (cf. Fig. 14). Arrived at maturity, it ceases to ingest

solid food-substances.   As old age comes on, it begins
to feed upon its green cells.   Groups of such cells in
all stages of digestion and varying in colour from
yellowish-green to brown may be seen lying in large
vacuoles in the central digestive tissue of the bodies
of old specimens of C. roscoffensis.   Thus, though,
as we shall see presently, the green cells of C. roscoffensis play an all-important part in the economy
of that organism, they are not the sole purveyors of
nourishment to it.   Throughout a considerable part
of its life, C. roscoffensis is able to help itself to the
solid food supplied by the micro-flora and fauna of
its environment.

Unlike C. roscoffensis, its ally, C. paradoxa, knows
no abstemious fits.   Throughout its life it is a glutton.
A glance at the body of the larval animal (Fig. 16)
gives the impression of a marine museum, so accom-
modating is the body of C. paradoxa.   There, may
be seen the remains of several scores of diatoms of all
shapes and sizes.   When examined immediately after
capture, a young or old C. paradoxa may be found to
contain, not only diatoms, but two or three Copepods,
each half as large as the animal itself, and, if it be late
in the summer, rows of tetraspores of red algæ show
through the transparent body like so many cardinal
buttons.   In C. paradoxa therefore, as in C. roscoffensis, though the coloured, plant-like cells may well
play an important and even indispensable part in the

life of the animal, they are not called upon to cater for all the food that it requires.

As a first step toward the investigation of their origin and *rôle*, we will make a microscopic examination of the coloured cells of our plant-animals.

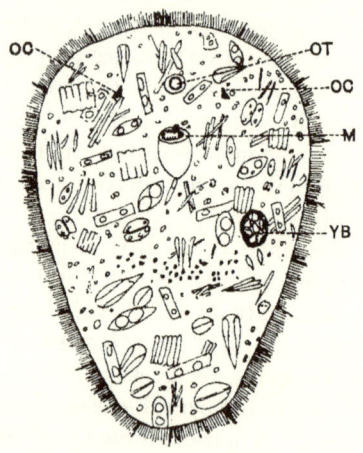

Fig. 16. A larval Convoluta paradoxa showing cilia and bristle-like projections from skin. YB = the only yellow-brown cell contained in the body. M = mouth and gullet. OC = eye spots. OT = otocyst.

The tissue of the body is crowded with large numbers of diatoms which have been ingested.

When a living C. roscoffensis is examined under the low power of the microscope, its green cells are seen to be, some spherical, some pear-shaped and

some of irregular form. As the animal moves along,
its muscles contract and the shapes of the green cells
change somewhat (cf. Fig. 14). Seen under a higher
power, the green cells present the appearances indicated in Fig. 17. Each cell or protoplast is made
up of a large green, and a small colourless part.
The former consists of the chloroplast, the latter of

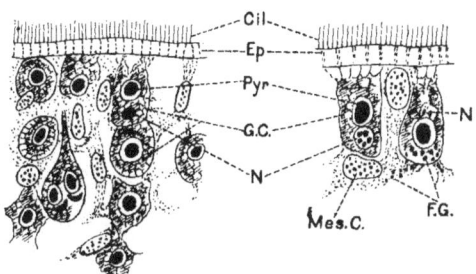

Fig. 17. Cross-section of the superficial tissues of Convoluta roscoffensis. G.C. = green cells in rows. N = nucleus of green cell. Pyr =
pyrenoid. Mes.C. = nucleus of attendant cell. F.G. = fat granules.
Cil = cilia at the surface of the animal. Ep = epidermis.

colourless protoplasm. Embedded in the mass of
colourless protoplasm, but not visible without special
methods of preparation, is a denser, oval body, the
nucleus which is an integral part of plant and animal
cells. Lying in the chloroplast is a dense body surrounded, halo-like, by a clearer margin. This body,
which is called a pyrenoid, consists of proteins and is

characteristic of the cells of many of the lower algæ (Fig. 21, p. 123).

If a green cell is treated with a solution of iodine, the nucleus and the pyrenoid are stained brown, and round the latter a thin, blue, granular layer may be distinguished. This layer is known as the starch sheath. As the action of the iodine continues, minute, lens-shaped starch grains—distinguished by their blue colour—may be seen lying in the chloroplast. Unlike algal cells in general, the green cells in the body of C. roscoffensis have no cellulose wall, but are bounded each by an elastic layer of protoplasm.

The yellow-brown cells of C. paradoxa are built on somewhat different lines. Each cell contains a number of irregularly oval, or polygonal, yellow-brown, discoidal chloroplasts which occupy about half of the cell (Fig. 18). The other half of the yellow-brown cell consists of clear, transparent, vacuolated protoplasm. By suitable treatment, involving the dissolution of the pigment, a nucleus may be made out, slung in the centre of the cell by threads of protoplasm which stretch from the periphery. When the cells are treated with alcohol, the yellow-brown pigment is dissolved away and green chlorophyll, previously screened by the yellow-brown pigment, is seen in the chloroplasts. The reaction is useful in that it enables us to distinguish the chloroplasts of the yellow-brown cells from the orange-coloured glands which occur in the surface-tissues of the body of the animal.

The green pigment of C. roscoffensis is chlorophyll, identical in its spectroscopic properties with that contained in the green tissues of plants. Moreover,

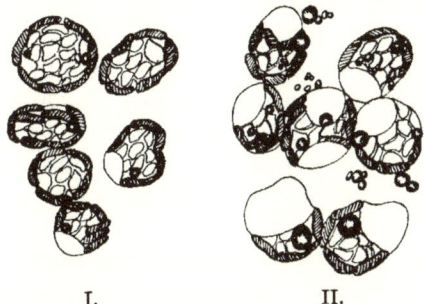

I.   II.

Fig. 18. Yellow-brown cells of Convoluta paradoxa. I. As seen in an animal some hours after capture. II. As seen immediately after capture. The spherical masses lying in the cells and also outside them represent the fat-globules referred to on pp. 89 and 91.

The shaded oval bodies at the periphery of the cells represent chloroplasts.

as Geddes (1879) has demonstrated, the green cells of C. roscoffensis are capable of photosynthesis. When the animals are exposed to light, they decompose carbon-dioxide, give off oxygen, and manufacture carbohydrates, the excess of which is stored in the chloroplast in the form of starch.

That the starch which occurs in the green cells

of C. roscoffensis owes its origin to photosynthesis, we demonstrate by the method which is used for a similar purpose in the case of plants. The living animals are kept in darkness and examined daily for starch. After a time—about seven or eight days in young C. roscoffensis, about fourteen days in older animals—when, as indicated by the samples tested, starch has disappeared—having been converted into sugar and used as food or in respiration—the animals are brought into the light and tested at intervals for starch. As is the case with green plants treated similarly, photosynthesis is resumed as soon as light falls on the green cells, and within less than ten minutes starch, which represents the reserve form of the photosynthesised carbohydrate, makes its appearance in the green cells. Moreover, the light which is most efficient for photosynthesis in plants, that of the red end of the spectrum, is also most efficient for photosynthesis in the green cells of the plant-animal.

It is not so easy to obtain rigid proof that the yellow-brown cells of C. paradoxa are capable of photosynthesis. Nevertheless, the indirect evidence supports strongly the view that they do actually function in this manner.

In the first place, like similarly coloured algæ, which are known to manufacture their food photosynthetically, they possess a screening pigment and also chlorophyll.

In the second place, when examined *immediately* after capture, the transparent reticulum of the cells (Fig. 18, II) is found to contain colourless, refractive globules or droplets, which, when treated with suitable reagents (osmic acid, etc.), may be recognised to consist of, or at all events to contain, fat. Now, it is well known that certain plants, some algæ among others, store their reserve, photosynthesised carbon-compounds, not as starch, but as oil. That these globules are of the nature of reserve substances derived from the products of photosynthetic activity is rendered probable by the following facts. First, when a catch of animals is divided into two lots, and one is kept in darkness and the other in the light, the reserve fat-globules disappear more quickly from the yellow-brown cells of the dark-kept animals than from those of the animals kept in the light. Second, if two similar batches of animals are kept in darkness, one in pure (filtered) sea-water, the other in sea-water containing sea-weed from the C. paradoxa zone, fat disappears from both, but more quickly from the yellow-brown cells of the starved animals. If the fat were derived from the food (sea-weed with its micro-flora and fauna) there would seem to be no reason why it should disappear at all from the yellow-brown cells of the fed animals. On the other hand, assuming that the fat-globules serve as food-material, not only for the yellow-brown cells but also for those of the animals, we should expect that the latter, when deprived of other supplies, would make larger demands on the reserve fat of the yellow-brown cells than when the animals had access to other food supplies. Third, if C. paradoxa are kept in filtered sea-water, and hence deprived of all food except that which it can obtain from the yellow-brown cells, then, so long as they are exposed to the light, the yellow-brown cells continue to contain fat-globules. Since animals deprived of food get just as hungry in the light as in darkness, it would appear to follow that the reason why the fat does not disappear from the yellow-brown cells of the light-kept animals is that, as fast as it is removed to serve for the nutrition of the animal, it is reformed by the yellow-brown cells. It is therefore to be concluded that the fat-globules are reserve products of the photosynthetic activity of the yellow-brown cells.

Thus we reach a definite stage in the course of our enquiry into the significance of the green and yellow-brown cells of C. roscoffensis and C. paradoxa. These cells are capable, in the same way as the chlorophyll-containing cells of plants, of manufacturing organic, carbon-containing substances from inorganic materials and of storing the surplus in the form of starch or fat.

Our next step must be to determine whether the products of the photosynthetic activity of the coloured cells are available for the nutrition of the tissues of the animals which contain them.

The evidence which suffices to demonstrate that the coloured cells do actually make contributions to the nutrition of the animals is not far to seek. If C. paradoxa is examined microscopically immediately after capture, it is seen that the tissues of animals whose yellow-brown cells are rich in droplets of reserve fat contain also large numbers of globules of a similar nature (Fig. 18). Moreover, the appearance of the fat-globules contained in the yellow-brown cells suggests most forcibly that the fat lying in the tissues of the animal owes its origin to the secretion of fat by the yellow-brown cells. The appearance of the yellow-brown cells recalls, in this respect, that of cells of a mammary gland in its active stage. Just as the fat contained in the milk which is secreted by the cells of a mammary gland is liberated in droplets by the rupture of the clear vacuolated parts of the secreting cells, so droplets may be seen in course of extrusion from the yellow-brown cells into the tissues of the animals (Fig. 18). The large, clear, anterior end of the yellow-brown cell—only to be seen in fresh-caught animals which have been exposed in their natural habitat to a fairly high light intensity—contains often one, large fat-globule. In some yellow-brown cells, one or more droplets lie in the deeper part of the clear anterior end, whilst, in others, a single, large drop lies close against the anterior margin of the cell, separated from the tissues of the animal only by the thinnest of membranes. Finally, other large globules may be seen lying just outside the colourless borders of yellow-brown cells and presenting the appearance of having been extruded from them. We conclude that the fat-globules, formed in the yellow-brown cells of C. paradoxa, pass by a process of secretion from these cells to those of the animal and serve the animal for nutritive purposes.

It is very probable that a similar secretion occurs in C. roscoffensis. For, in the first place, starch, which, as we have learned, appears in the green cells as the result of photosynthesis, does not occur in the other tissues of C. roscoffensis. In the second place, this animal does not possess the power of digesting starch. When supplied with starch grains, it ingests them readily, transfers them to the vacuoles which lie in its digestive tract, but is unable to dissolve them. They remain for a time in the vacuoles, and are then discharged by a temporary rupture of the surface of the body. In the third place, in carefully prepared and stained sections through the body of C. roscoffensis, there may be seen rows of fatty granules passing from the green cells to the neighbouring animal cells (Fig. 17, *F.G.*). Nor does a conversion of starch into fat present any difficulty to vegetable cells. For example, in many trees, the tissues of the trunks contain, in autumn, large stores of starch ; as winter advances the starch is replaced by oil or fat, and, again, when spring arrives, the oil is reconverted into starch. It is of course open to us to suppose that, just as the sugar formed photosynthetically by the green cells of a leaf is translocated as fast as may be through the tissues of the leaf-stalk and stem to meet the demands of the colourless cells of the plant which depend for their food supplies on the activity of the green cells, and just as the starch which appears in the green cells of the leaf represents only the surplus which is stored temporarily in a convenient form to be changed to sugar and distributed later ôn ; so, in C. roscoffensis, the photosynthesised sugar streams away as such to the colourless cells of the animal, only the surplus being stored as starch.

Whether it travels as sugar or, as the former observations seem to indicate, as fat, there is no doubt that the organic, carbon-containing food-material, produced photosynthetically by the green cells of C. roscoffensis, serves for the nutrition of the animal's tissues.

Indeed, as we show presently, unless the green cells are present in the body of the animal, and unless they increase and multiply therein, the animal does not grow at all.

The phase in the relation between coloured, chlorophyll-containing cells and animal tissues which we have just described, presents the closest parallel with the relation which obtains between the green and non-green cells of any chlorophyllous plant. In both plant and plant-animals, the chlorophyll-containing cells manufacture carbohydrates in excess of their own requirements, and, in both, the excess is translocated to the colourless tissues and used by them as food-material.

But, in certain circumstances, C. paradoxa and, to a somewhat less degree, C. roscoffensis may exploit their coloured cells in a more summary manner.

Thus, when animals are kept in darkness in sea-water filtered through a Pasteur-Chamberland filter they become reduced greatly in size. The reduction in size is, as we know, greater, and takes place more rapidly, in dark-kept than in light-kept animals. In one experiment in which the animals were measured, those which had been kept in darkness were, on the average, two and a half times as small as those which had been kept in the light; the average superficial dimensions of the dark-kept C. paradoxa being ·08 square inch and those of the light-kept animals ·2 square inch.

The powers of resistance to starvation of both C. roscoffensis and C. paradoxa are extraordinary. Thus, it is possible to maintain C. paradoxa alive for upwards of a month in filtered water ; that is, under conditions, in which it is deprived of all external supplies of food. When subjected for long periods to these conditions, the animals become reduced in size and—as is the case to a yet more marked degree with those kept in darkness—also show an extraordinary reduction both in number and size of their yellow-brown cells.

In prolonged darkness, the yellow-brown cells, once their reserves of food-material have been extracted from them to meet the needs of the animal, are digested wholesale by C. paradoxa. If the water in which they are contained is altogether devoid of food supplies, the attack by the animal on its coloured cells occurs all the sooner. Even in the light, if external food supplies are withheld from C. paradoxa, a time comes when, although the yellow-brown cells are supplying it with photosynthesised food-materials as fast as they can under the difficult circumstances, it turns upon them ;—killing and digesting the goose which laid its golden eggs.

Microscopic examination of animals kept in prolonged darkness supplies evidence that the degeneration of the yellow-brown cells is not a mere decay within the body, but is the result of a true process of digestion exerted on them by the animal. The first sign of digestive action is a reduction in size of the yellow-brown cells. They assume a more spherical shape and their chloroplasts become smaller and rounder. Each reduced algal cell is now seen to lie in a distinct, digestive vacuole containing a pink fluid. Next, the pigment of the chloroplasts is dissolved and, diffusing out of the cell, may impart a brown colour to the vacuolar fluid. At this stage, the chloroplasts are greenish; later, they become colourless. Finally, heaps of few or many, colourless, curiously persistent granules are all that remain of the algal cells.

It is interesting to observe, in this connection, that if animals are brought, after a prolonged sojourn in darkness, into the light and supplied with fresh sea-water, yellow-brown cells make their appearance again in their bodies. As they grow and increase in numbers, the animals also begin again to grow.

So also in the case of C. roscoffensis, if the green cells fail to make their appearance in the body, the animals remain of microscopic size. If, on the other hand, the green cells appear, increase and multiply to form the characteristic green tissue, the animals begin to grow rapidly.

Thus in various ways it has been demonstrated that C. roscoffensis and C. paradoxa depend for their food on their coloured cells. Without them, they fail to grow. When, by exposure to darkness, the coloured cells are put out of photosynthetic action, the animals become reduced in size, and, after giving their coloured cells a respite of some weeks, they turn on these algal cells and digest them. In C. paradoxa, this raiding of the coloured cells occurs only under special, artificial conditions ; as, for instance, during prolonged darkness. But, in the case of C. roscoffensis, it is a regular procedure with animals which have reached a certain age. Nor is the reason for this difference of behaviour between the two plant-animals far to seek. Whereas C. paradoxa retains its habit of ingesting solid food and looks to its yellow-brown cells for supplementary supplies only, C. roscoffensis, at a certain stage, shuts its mouth and cultivates its garden of green cells. Now, inasmuch as hunger—cell-hunger —may be due to one or more of many different lacks, lack of carbohydrate, lack of nitrogenous food-substances, or of mineral compounds, it is bound to happen sooner or later that the animal part of C. roscoffensis, in its phase of total abstinence from food, will feel the pinch of one kind of hunger or another. Goaded by this all-powerful stimulus it turns upon its green cells, and, biting the hand that fed it, seeks, by devouring them, to satisfy its cravings for some special food-substances.

To the question what particular kind of hunger is it that drives the animal to devour its plant-like cells, we shall address ourselves, after we have investigated, in the next chapter, the origin of these cells. For the present, we content ourselves with summing up what we have learned of the relations between the animals and their plant-like cells.

The coloured cells manufacture photosynthetically food-materials, storing the surplus as starch and fat. The animal receives from the coloured cells supplies of food-material. So plentiful are these supplies in C. roscoffensis that the animal comes, in course of time, to rely altogether upon them for its nutrition. Ceasing to take up food, it grows, bears eggs, and produces young at the expense of the materials supplied by its green cells. This life of curious asceticism leads, however, to trouble. Though the green cells continue to supply organic carbon compounds, something or other is lacking from the prepared food which the animal thus receives. To make up for this lack, it digests in detail its green cells, coming often in old age to present a strange appearance—head-end green, tail-end white. Having exhausted its stores of green cells, without apparently satisfying all its needs, it pines away and dies.

In its earlier youth, C. roscoffensis feeds, after the manner of animals in general, on other plants or animals. This is the first phase. In the course thereof, green cells appear in the body, increase, multiply, photosynthesise and distribute food materials to the animal's tissues. For a while, C. roscoffensis receives food from two sources—from ingested plants and animals and from its green cells.

This second phase is succeeded by a third, in which C. roscoffensis, having ceased to ingest solid food, is nourished, in the same manner as the colourless non-chlorophyllous tissues of a green plant are nourished, by the products of the photosynthetic activity of its green cells.

Last stage of all which ends this strange eventful history:—the animal digests its green cells, and, having done so, dies.

In the first phase, the mode of nutrition is animal-wise : in the second, part animal-, part plant-wise : in the third, altogether plant-wise or holophytic : and in the fourth, autotrophic, that is by living on itself.

Convoluta paradoxa is like unto C. roscoffensis, except that its experiments in nutrition stop, under normal circumstances, at phase two. Under artificial conditions, however, it behaves like its ally, lives for a while like a plant at the expense of the products of photosynthesis of its yellow-brown cells, and, finally, driven to digest these cells, prolongs its life autotrophically.

CHAPTER IV

THE ORIGIN AND NATURE OF THE GREEN CELLS OF CONVOLUTA ROSCOFFENSIS.

GREEN, yellow or brown cells, resembling in a general way those contained in the bodies of C. roscoffensis and C. paradoxa, are found in many different kinds of animals belonging to the lower groups of the animal kingdom.

Such cells are known to exist in representatives of every division of the free-living Protozoa—the lowest group of animals. They occur in certain sponges, in many sea-anemones and in various species of coral-forming animals. In higher groups, they are rare though they are known to occur in isolated cases, for example, in Zoobothrium, a member of the Polyzoa, in Elysia (a Mollusc), and in Echinocardium (an Echinoderm).

The best known example of an animal containing green alga-like cells is the common, freshwater hydra, Hydra viridis.

In certain of the animals which are characterised by the possession of coloured cells, these peculiar elements are invariably present. In other animals, the coloured cells may occur in some individuals, but not in others. The former, general association we may call obligate, and the latter, occasional association, facultative.

Hydra viridis, Convoluta roscoffensis and C. paradoxa are examples of organisms in which the association is obligate.

Facultative association may take one of two forms. Either some specimens living in a given region may possess coloured cells, whilst other specimens of the same region lack them, or a given species may consist, in one part of its range, of individuals all of which contain coloured cells, and, in another part of its range, of individuals none of which possess them. For example, Noctiluca is colourless in the North Atlantic, but green in the Indian Ocean. British Alcyonium have no chlorophyll-containing cells, whereas the nearly allied Alcyonium ceylonicum possesses them. It seems probable—and this is a point of which we shall make use presently—that association between animal and plant-like cells is commoner in the warmer than in the colder seas.

The problem of the origin and nature of the green, yellow and brown cells which occur in animals has engaged, from time to time, the attention of zoologists. Long ago the name Zoochlorella was given to the green cell and Zooxanthella to the brown or yellow-brown

cell. Since, however, these names are applied, respectively, to any green and any brown plant-like
cell which occur in any animal their value is but
limited.

That Zoochlorellæ and Zooxanthellæ are plant-
like cells is undisputed. They contain chlorophyll,
decompose carbon-dioxide with evolution of oxygen,
may, in the case of Zoochlorellæ, contain starch: a
substance for the manufacture of which plants and
not animals possess the secret. Further, in some
cases, at all events, the coloured cells possess a wall
of cellulose, another substance the formation of
which is confined exclusively or almost exclusively
to members of the vegetable kingdom.

Beside one or more chloroplasts, a nucleus and
a pyrenoid, the coloured cells have been shown in
some cases to contain a small, bright red body known
as an eye-spot (Fig. 21, p. 123). In free-living, unicellular algæ, the eye-spot serves the purpose of light-
perception and thus is part of the nervous machinery
for the performance of phototropic movements. Hence
its occurrence in green cells imprisoned in the bodies
of animals may be regarded as a strong indication
that the green cell which possesses it had once a
free-living existence.

Nevertheless, though such facts as these lend
powerful support to the hypothesis that Zoochlorellæ
and Zooxanthellæ are algal cells which have aban-

doned their free and independent modes of life and
have taken up their abodes in the tissues of animals,
yet they do not constitute a final proof of the truth
of this hypothesis. Indeed, the problems presented
by the chlorophyllous cells of animals are too
numerous and important to be dismissed by means
of a loosely-drawn inference of this sort. To the
possession of chlorophyll the plant owes its powers
of photosynthetic manufacture; and to the absence of
this pigment from the cells of animals is due the
dependence of the animal world on the world of
plants for food supplies. Yet, low down in the
animal kingdom, organisms exist which, though undoubtedly possessed of distinct animal characteristics,
contain chlorophyll and use it for the manufacture
of carbohydrate food. Thus, species of Euglena
(e.g. E. viridis), which stand near the parting of the
ways which lead, the one to the animal kingdom,
the other to the vegetable kingdom, contain chlorophyll and use it for photosynthetic purposes. Now
Euglena viridis is undoubtedly an animal. The single
cell or protoplast of which it consists is provided
with a gullet, into which solid particles may pass
and thus be ingested by the animal. The membrane
which encloses the organism is not composed of
cellulose—the cell-wall substance of typically vegetable organisms; and in yet other ways Euglena
gives evidence of its "animal" nature.

Although zoologists and botanists are agreed that the genus Euglena belongs to the animal kingdom, yet it possesses the power of constructing a green pigment—chlorophyll—which is identical in physical properties with that which occurs in the chloroplasts of plants. Here there is no question, apparently, of any swallowing by Euglena of plant cells. The animal cell makes the pigment in the same way as a plant cell makes it, and, having made it, uses it for photosynthetic purposes.

In certain circumstances, chlorophyll disappears from the body and Euglena viridis passes into a colourless phase. When in this state the animal, if it is to feed at all, must do so by ingesting ready-made food. That is, from being a holophytic organism—one with a typically plant-like mode of nutrition—it becomes heterotrophic, that is, it feeds on ready-made, organic materials, obtained from its environment. After a time, it may reconstruct its chlorophyll and become free once more to manufacture by photosynthesis its organic food-substances from the raw, inorganic materials of its environment.

If one species of animal can do this, why should not other, even more highly developed species, possess like powers? Why should there not appear, here and there, animals which resume the habit possessed by their ancestors, construct chlorophyll and become independent, photosynthesising organisms?

Or, to pursue another line of argument. The Zoochlorellæ of some animals are typically plant-like cells. They possess a chloroplast, a nucleus and a cellulose cell-wall. But in other animals, in C. roscoffensis, for example, the green bodies are of simpler build. Each consists of a naked protoplast which is made up of a green chloroplast and a colourless mass of eccentrically lying protoplasm in which a nucleus may be included (Fig. 17, p. 85). The green tissue, composed of vast numbers of these elements, appears to be as much a part of the animal as any other of its tissues. So much is this the case that all attempts to cultivate the green cells of C. roscoffensis outside the body end in failure. They are no more capable of independent existence than are the chloroplasts of the chlorophyllous tissues of a green plant.

What is there to prevent us from assuming, as Haberlandt has assumed, that the green cells of C. roscoffensis are not complete cells but merely chloroplasts, and that, like the chloroplasts of the green plant, they are transmitted as colourless particles (leucoplasts) from the organism to its eggs, and, multiplying as the egg divides to form the embryo, reappear as green chloroplasts in the tissues of the new generation? On this hypothesis the colourless part of the green cell of C. roscoffensis (Fig. 17) is an animal cell which attends upon the green chloroplast. In other words, just as a green cell of a flowering plant

consists of colourless, nucleated protoplasm containing chloroplasts, so, on Haberlandt's hypothesis, the
green cell of C. roscoffensis consists of a colourless,
animal part containing a green chloroplast.

Pursuing this hypothesis to its natural conclusion,
it is easy to imagine, with Haberlandt, that, in some
remote past, algal cells came to exist in symbiosis
with colourless C. roscoffensis; that the animal offered
such a congenial lodging as to induce the algæ to give
up going out altogether. They abandoned their cell-
wall as an enclosing apparatus no longer of service
to them. In return for security and all the comforts
of a home the green cells prepared the food both for
themselves and for their host. Submitting itself to
the guidance of the animal, the green cell abandoned its nucleus and became reduced to a naked
chloroplast.

So it might have come about that the only powers
retained by this relic of a once complete and free
algal cell are those possessed by the chloroplast of
a green plant, namely, the powers of photosynthesis
and of division to form new chloroplasts. Moreover,
just as the chloroplasts contained in the egg-cells of
plants lose their green pigment, and become colourless
leucoplasts, which, dividing as the cells of the plant-
embryo divide, give rise to the chloroplasts of the
next generation, so, on this hypothesis, it would follow
that the green chloroplasts of C. roscoffensis might

give rise to colourless leucoplasts which pass into
the egg and provide the rudiments from which the
chloroplasts of the larval animal are developed.

Yet again, if this were indeed the course of events
in C. roscoffensis, if, from their free, complete condition, the original green cells which gained access
to the body of our plant-animal have become reduced
to mere chloroplasts, might not this animal provide an
illustration of the mode of origin of the higher green
plants themselves? In a remote past, a symbiosis
was struck up between a colourless organism and
a green alga—such a communal mode of life, for
example, as that presented by lichens at the present
day.   Convenient models these to show us the
relation between colourless organism and undoubted
algal cells.   So happy is the hypothetical partnership
between alga and colourless organism that new
developments ensue.   A new and composite form of
life comes into existence.   The colourless tissues
burrow in the earth and supply, along well-defined
conduits, the water and minerals required by the
green cells.   They form tall trunks and spreading
branches to lift the chloroplasts—the representatives
of the algal cells—nearer to the sun. The green plant
is in being.

Of this alluring picture, evoked by syren-voiced
hypothesis, we are bound to ask the simple, sober
question, is it true? To this question we can give

no answer until we have discovered experimentally
the origin of the plant-like cells occurring in each
species of the many animals which possess them.

This we proceed to do in the case of C. roscoffensis.
Two methods are open to us for the purpose. We must
either trace back the green cells to the earliest stage
at which they make their appearance in the animal
and ascertain whether they may be then identified
with any known, free-living alga. If we succeed in
this, we shall have obtained, not absolute proof, but
strong ground for believing that the green cells are
of intrusive origin. Or—and this is the only certain
way—we must cultivate the alga, and having obtained animals which are free from it, and having
demonstrated that such animals remain indefinitely
colourless, we must infect the animal with the algæ
of our pure algal-culture and synthesise the green
plant-animal.

As we have indicated already, all attempts to
isolate living green cells from the body of C. roscoffensis have failed ; and so it would seem that the
former, less satisfactory method alone remains. The
application of the method is simple enough. It
consists in the microscopic examination of larval
C. roscoffensis in all stages, from the time of hatching up to the time when green cells, resembling those
of the adult, may be recognised within the body.

When just-hatched C. roscoffensis are examined with the high power of the microscope, no green cells are to be seen in their bodies, nor are there present any colourless cells resembling in shape or structure the green cells, nor do either eggs or larvæ appear to contain leucoplasts.

If just-hatched animals are transferred to sea-water filtered by means of a Pasteur-Chamberland filter, though, in the course of time, some may become green, many remain colourless. Therefore it is highly probable that the green cells do not owe their origin to colourless antecedents (leucoplasts) present in the eggs. For, were such forerunners of the green chloroplasts present, they would develop into chloroplasts in all, or at all events in the great majority, of the larvæ. On the other hand, if young animals are hatched and allowed to remain in ordinary unfiltered sea-water, green cells make their appearance with certainty in the animals in the course of one or two days. In the youngest larvæ, there are to be seen no more than two or four green cells, each of them lying in a clear vacuole and occupying a fairly definite situation in the body. Two such cells lie right and left, a little behind the otocyst and two right and left about the middle of the body. By their repeated division is produced ultimately the whole contingent of green cells of the adult body.

At stages earlier than this, no green cells are to

be found, but a larger or smaller, colourless body may
be seen lying in a central vacuole in such a situation
as to suggest that it has been taken up through the
mouth.   The larger body consists of two closely
opposed cells, the smaller of a single cell.   In
either case, the colourless body is surrounded by a
mucilaginous wall which swells considerably as its
contents divide, in the case of the single cell into
four, in that of the large cell, into eight daughter
cells.   The colourless cells, each about 15 to 16 μ in
length ($=$ about $\frac{1}{1600}$ inch), are discharged by the
bursting of the vacuole and take up positions similar
to those in which the four green cells are found.
Though colourless and of granular content, a large,
oily looking pyrenoid may be made out in each cell (cf.
Fig. 22, B, $Pyr.$, p. 125), and by appropriate methods
of staining, the presence of a nucleus may be demonstrated.   The colourless cells increase in size and, in
each, a red eye-spot makes its appearance as a little,
lateral patch near the margin.   Soon a distinction is
to be seen between a colourless plug of protoplasm
and a cup-shaped, granular mass which occupies the
major part of the cell (Fig. 22 C).   A faint yellow
colour steals over the cup-shaped, granular mass, the
yellow colour deepens and, becoming green, enables
us to identify the cup-shaped, granular mass as a
chloroplast.   The cell, now a green cell, possesses no
cell-wall, and differs only from a green cell of an adult

C. roscoffensis in its more regular, oval shape and in the possession of an eye-spot. Each green cell divides. The daughter cells formed by the division lack the oval shapes of the mother cell: they lack also eye-spots. The colourless plug or neck of protoplasm no longer occupies the position of a cork in a flask, but lies eccentrically to the chloroplast and in it the cell-nucleus is contained. In short, they are identical with the green cells of the adult animal. Thus we reach two conclusions of importance. First, that the green cells of C. roscoffensis are preceded by colour-less cells. Second, that the mass of colourless proto-plasm attached to the green cell is not, as Haberlandt suggests, an animal cell standing in close relation with a chloroplast, but is an integral part of the green cell. As the young green cell continues to divide, a significant change may be observed in the shape and state of the nucleus. Distinct and spheri-cal in the colourless and original green cells, it becomes, in the cells produced by successive divisions, more granular and indistinct, till, when the number of green cells has increased considerably, some only among them may be seen to contain nuclear material —fine granules in a clear area: the rest contain no trace of nuclear material. In other words, the great majority of the green cells of the adult animal are not complete cells, but cells which show all stages of diminishing nuclear substance (Fig. 17). Inasmuch as it is a well-established fact that the nuclear part of the protoplasm plays an important *rôle* in the life and work of the cell, these observations throw light on the subordination of the green-celled tissue of C. roscoffensis to that of the animal. Since, also, the nucleus is known to play a part in cell-wall formation, we are no longer surprised that a cell-wall fails to form in the green cells. Further, in this progressive nuclear degeneration, we have the explanation of the inability of the green cells to survive separation from the tissues of the animal. Those green cells whose nuclei are least degenerate are capable of division, but even they have suffered. They are no longer able to form a cell-wall nor to exist as independent organisms. As division succeeds division, the nuclear material becomes further reduced till, in the adult animal, it is often difficult to find any sign of nucleus in the large majority of the green cells.

It is highly probable that the advent of this enucleate stage in the green cells is the signal to the animal to devour them. Though still capable of photosynthesising, the green cells, unable to offer resistance to those of the animal, are surrounded by the latter, devoured and digested. A significant phenomenon is revealed by the drawings (Fig. 17) of sections through the green cells of C. roscoffensis. In places, ingrowing rows of cells may be seen budded off from the outermost green cell and, of these rows, only the outermost contain a distinct nucleus, others possess deep-staining granules, and others no nuclear material whatever. A parallel suggests itself between the green cells of C. roscoffensis and the red blood-corpuscles of the higher vertebrates. As the red discs are enucleate, partial cells budded off from the nucleated red cells, so may the green cells be regarded as enucleate, partial cells budded off from the outermost, nucleated green cells ; and, as the red blood corpuscles are of limited life and specialised (respiratory) function, so are the green cells of C. roscoffensis of limited life and of specialised (photosynthetic) function.

The green cell, devoid of nucleus, would not, however, appear to be shut off from all nuclear influences. For the enucleate green cell may be connected by fine processes with another green cell still possessed of nuclear substance (Fig. 17, p. 85). Moreover, such green cells as are without nuclear material are accompanied by a large attendant nucleus of animal origin. This close association of "attendant nucleus" and green cell is shown in Fig. 17, *Mes.C.* It may be that the attendant nuclei are those of "wandering cells" of the animal which lie in wait for enucleate green cells and, at a subsequent stage, digest them.

The astonishing closeness of the relationship between animal and green cells offers some support

for the hypothesis, suggested independently by
Schimper and Lankester, which we have already
outlined as to the composite nature of higher green
plants. The algal cells of C. roscoffensis are on
the road which leads to complete loss of independence. In the higher green plant this loss is
complete. The green cells of C. roscoffensis lose
cell-wall and nucleus, but retain some colourless
protoplasm; the green elements of the flowering
plant—if they are regarded as the descendants of
originally free algæ—have lost everything except the
photosynthesising organs—the chloroplasts.

But a wide gap remains between the state of
affairs in C. roscoffensis and that in the higher green
plant. Sooner or later C. roscoffensis destroys and
digests its green cells, and none of them, nor any
colourless representatives of the green cells, pass into
the egg-cells; whereas the higher green plants provide for the future crop of chloroplasts in their
descendants by transmitting colourless rudiments of
the chloroplasts to their egg-cells.

An adult C. roscoffensis is a complex of two
organisms—one, the colourless animal, the other, the
chloroplast-remainders of the original, green, nucleated, algal cells. In its case, unlike that just imagined
for the green plant, the synthesis is not a permanent
one. It endures but for the lifetime of the animal
and has to be recommenced in every larval Convoluta.

The discovery that the green cells of C. roscoffensis arise, in the larval animal, from a colourless cell which lies in a vacuole near the mouth of the animal, makes it all but certain that they are of extraneous origin, and that, in the course of ingesting solid food, the larvæ take up also these antecedents of the green cells.

Failing the isolation and cultivation of the green cells, and failing the discovery of the colourless or green cells in the sea-water, all that seemed possible to do more was to demonstrate that animals hatched in pure, filtered sea-water, remain colourless. The method adopted for this purpose was as follows. Large numbers of animals were scooped up in a watch-glass as free from sand as possible. They were brought into the laboratory, allowed to geotrope into a white cup and washed repeatedly with filtered sea-water. Their habit of sticking to the surface of the cup after it had been tapped gently, permitted of the water being poured off without any considerable loss of animals. After washing them many times, the animals were transferred to filtered sea-water in large glass dishes. In due season, the egg-capsules were formed and since, though very minute, they are visible readily to the practised eye, they could be picked out and transferred yet again to filtered sea-water. In this they were allowed to hatch. Though the results of such experiments, which were repeated many times, confirmed the infection-hypothesis they were not sufficiently uniform to establish it absolutely. Time after time, the minute larvæ showed, when examined microscopically, a general absence both of green cells and colourless precursors of green cells; but, time after time, also, an occasional animal reared under these apparently sterile conditions was found to contain green cells. Though the occurrence of green cells among animals hatched in filtered sea-water was infrequent and sporadic, yet there such animals were and, to make matters worse, the longer the larvæ were kept under observation, the larger was the number of specimens which contained green cells.

Evidently, either the infection-hypothesis was wrong or there was some defect in the experiment. A careful examination of the conditions of the experiment revealed ultimately a source of error and opened up the possibility of a new origin for the green cells. The mucilaginous capsules, enclosing groups of eggs, were discovered to be infested with all sorts of minute organisms. On the capsules, and in them, were many different kinds of microscopic animals and plants—diatoms, infusoria and forms of life unlike any to be seen elsewhere. Since the egg-capsule is formed by a secretion of the skin, and since the skin of the animal is covered with slime, it was at once clear that repeated washing in filtered sea-water

had not succeeded in making the animals biologically
clean. It was clear, also, that, if the infecting organism
came from the egg-capsule, it might be derived not
from the sea-water but from the body of the parent.
At the time of hatching, there might be liberated
from the body of the animal, colourless or green
cells which, though they could not live in sea-water
and were not to be cultivated artificially, might well
be capable of living in the walls of, or inside, the
egg-capsules.

Fortunately, the elimination of this source of
error though laborious is not impossible.   When
ready to hatch, a very little help suffices to enable
the young to escape, not only from the thin membrane which encloses each, but also from the common,
mucilaginous egg-capsule.   Thus, by drawing a clutch
into a small pipette and then ejecting it and the water
from the pipette, the capsule bursts and the young
escape.   In this way, it was possible to separate
larvæ from their capsule-remnants.   By employing
this method, large numbers of larvæ, white, minute
and active, were isolated in filtered sea-water in
which the only source of infection lay in such invisible shreds of the capsule as might have happened
to get themselves transferred with the larvæ to the
filtered sea-water.

The method proved successful.   In one case, out of
forty-four larvæ isolated from capsule-remnants and

kept under observation for nearly three weeks, only five animals contained green cells. In another case, not a single animal of a total of forty-seven was found to contain any trace of green cell or colourless precursor.

Further, on transferring such uninfected animals to ordinary, unfiltered sea-water, they became uniformly green in the course of one or two days.

We conclude therefore that the green cells of C. roscoffensis are algæ; that the species to which they belong exists as a free-living, independent, marine plant; that this alga has a colourless stage, as well as a green stage, in its life-history; that the alga lives on the egg-capsule as well as in sea-water; that it is ingested with the food, and, resisting digestion, is planted in the body where it increases and multiplies and forms the green tissue of adult C. roscoffensis.

The questions remain: What is this alga and what does it look like in its free stage?

All sorts of attempts, some ludicrous in their extravagance, were made to isolate the infecting organism; whilst all the time it was lying under the eye. None of the attempts succeeded, and, during the winter, when experimental work could not be carried on, there was nothing to do but to contemplate ruefully the note-books recording the failures. But some wise person once observed, " You learn to play cricket

during the winter and to skate during the summer ";
—a paradox which contains a biological truth, for
the effects of exercise sum themselves up and write
the addition in our experience, not during the exercises, but in the intervals between them. At all
events, it was in the winter that a scrutiny of results
of the previous summer's work showed that when
green animals appeared among larvæ left with their
capsules in filtered sea-water, the manner of their
appearance was like that in which an epidemic
declares itself. At first, it marks down a single
victim, then some of the neighbours are affected till,
by and by, the disease is general. So it was with
the green-cell infection of C. roscoffensis. For
days after animals hatched in sea-water had become
quite green, those hatched from capsules kept in
filtered sea-water remained colourless. Then one
green, among many non-green, appeared. The
numbers of green animals increased, till, finally,
all became green. At once the conclusion presents
itself. In our experiments, the colourless stage of
the larvæ, which lasts as long as fourteen days, is an
incubation period, not for the animal but for the
infecting organism. Here or there, in spite of many
washings, a single algal cell which had settled on
the surface of the body has remained entangled in
the slime covering the animal. Transferred during
egg-laying to the capsule, it grows and divides. It

increases till it forms a multitude. Then the capsule
bursts and swarms of infecting organisms are liberated
and, ingested eagerly by C. roscoffensis, give rise to
the chlorophyllous cells of its body. The idea suggests the simple method : isolate the empty capsules,
as well as the just-hatched young, and in a week or
two some of the capsules will be found teeming with
the infecting organism.

On returning to Brittany in the following summer
the first thing done was to test the hypothesis.
Animals were washed and put to lay in filtered
sea-water; the egg-capsules were washed likewise and,
when the larvæ were hatching out, the latter were
put in one vessel and the remains of their capsules
in another. The animals remained colourless, though,
when samples of them were put into ordinary sea-
water, green cells made their appearance in their
bodies with uniform regularity. After seventeen
days, several small, green, globular bodies, each as
large as a big pin's head, made their appearance in the
water of the vessel containing the capsule-remnants
(Fig. 19). Their hue was the dark spinach-green of
C. roscoffensis. On microscopic examination, under
the slight pressure of a cover-glass, the dark green
mass dissolved and formed a cloud of active, green
flagellated cells, emerging from an egg-capsule (Fig. 20).
Though these free cells differed in various details from
the green cells which occur in the body of C. roscoffen-

sis, it was evident at once that they represented a stage in the life history of the infecting organism.

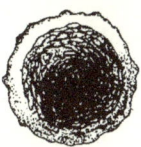

Fig. 19. Egg-capsule of Convoluta roscoffensis occupied by a dark mass consisting of vast numbers of the "infecting organism." (Magnified forty times.)

The final proof was applied. To the vessel containing colourless, uninfected Convolutas, some of the free, green cells were added. Within two days all

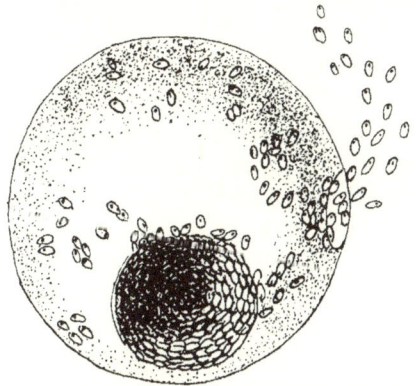

Fig. 20. The capsule shown in Fig. 19 enlarged and compressed during microscopic examination: the "infecting organism" escaping.

the animals were green. The synthesis of the plant-animal had been effected. As the result of introducing an undoubted green alga to a colourless, larval C. roscoffensis, a green plant-animal was formed.

Since a similar last stage has not been reached in the case of C. paradoxa—though to reach that stage is but a matter of time and experiment—we will devote ourselves not to a description of the incomplete evidence of its infection by a yellow-brown algal cell (see Keeble, 1908), but to a continuation of the study of the life-history of the infecting organism of C. roscoffensis. The securing of material for this purpose is comparatively easy once the art of cultivating the organism on the egg-capsules has been learned. It is facilitated also by the fact that the motile, green cells are, like C. roscoffensis, positively phototropic and assume, in a vessel of sea-water exposed to unilateral light, a position identical with that taken up by C. roscoffensis. Thus, though each algal cell is far too small to be visible to the unaided eye, the green, motile cells aggregated together at the surface of the water on the side toward the light become collectively visible as a green scum at or just above the water-line. The fact that they react to light by a vertically upward movement as well as by a movement toward the source of light suggests that the pyrenoid (Fig. 21), a dense blob of protein surrounded by protoplasm, may perform for the green

cell a function similar to that which the otocyst
performs for the animal.   Just as the granule of chalk
which is contained in the otocyst serves, by its gravitational movements, to stimulate C. roscoffensis to
orientate itself, so may the pyrenoid, falling now this
way and now that, serve to stimulate the protoplasm

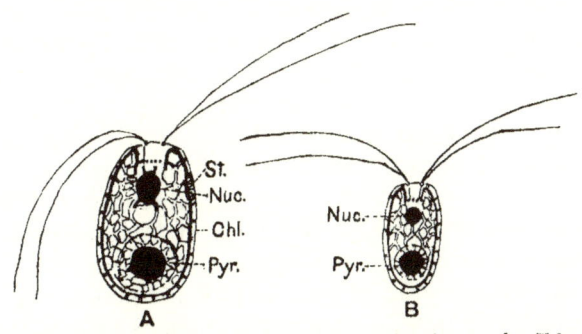

Fig. 21.   The infecting organism—an alga belonging to the Chlamydomonadineæ—seen under the high power of the microscope.
A = macrocyte. B = microcyte. Chl. = chloroplast (represented by
a reticulum) occupying the greater part of the cell. Nuc. =
nucleus. St. = eye-spot, indicated in A, but not in B, in which,
however, it occupies a similar position. Pyr. = pyrenoid. The
four long threadlike projections represent the flagella.

of the green cell to perform like movements of
orientation.

    If the green cells which have taken up their
position along the water-line are examined microscopically, they are found to include many which are
in active movement.   Each such cell resembles, in

essentials, the first green cells which appear in the body of C. roscoffensis. The bulk of the cell (Fig. 21) is occupied by a flask-shaped chloroplast (*Chl.*) in the middle of which lies the pyrenoid surrounded by its starch sheath. In the "neck" of the flask-shaped cell lies a colourless plug or core of protoplasm in which the nucleus is suspended. On one side of the chloroplast, a red eye-spot is placed (*St.*). So far, the description of the active cell corresponds exactly with that of one of the first green cells to be seen in the body of the infected animal. But, in addition to these structures, two others are met with in the free, active green cell which are absent from the green cells contained in the body of C. roscoffensis. These new structures are flagella and cell-wall.

The flagella (Fig. 21) consist of four equal, delicate protoplasmic threads each about twice as long as the green cell. They project from the anterior end of the colourless plug of protoplasm, and by their active, contractile movements serve to row the animal through the water. The cell-wall which invests the alga is extremely delicate and gives, when treated with appropriate reagents, the reaction not of cellulose but of chitin.

The flagellated cells are remarkable in that they occur in two sizes (Fig. 21). The large green cells—macrocytes—are about twice as big as the small microcytes. Such a difference in size occurs not infrequently among unicellular green algæ, and in cases where it occurs it has been shown that the macrocytes and microcytes fuse in pairs. As a result of this fusion, a single cell (or zygote) is produced, which, after passing through a period of rest, gives rise by division to four or more daughter cells. Since a similar fusion of two cells (or gametes) is the essential characteristic of sexual reproduction in all plants and animals, this fusion may be regarded as a process of sexual reproduction. Under ordinary circumstances, however, the macrocytes and

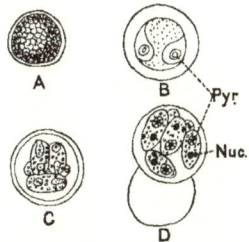

Fig. 22.   A = Resting cell (green or colourless) of the infecting organism.   B = Colourless cell dividing to form daughter cells. C = Resting cell containing four daughter cells.   D = Resting cell with eight daughter cells (six shown).   Nuc. = nucleus. Pyr. = Pyrenoid.

microcytes of the infecting organism do not fuse with one another.

Other cells taken from the green streak along the water-line possess no flagella (Fig. 22). They have settled down, withdrawn these structures and have surrounded themselves each with a thick wall of mucilage. So rapidly may the wall form about the encysting green cell, that individuals are sometimes observed in which the flagella may be seen in undulating movement within the enclosing wall.

The resting cells (Fig. 22) vary remarkably both in form and behaviour. Thus, a single, flagellated cell may come to rest, surround itself with a thin wall and divide longitudinally into two or four daughter cells. Each daughter cell, at first naked, organises a delicate cell-wall, develops flagella and escapes from the deliquescent mother wall as an active, flagellated cell. Or an active cell comes to rest, surrounds itself with a thick wall, takes on a spherical shape and becomes uniformly green (Fig. 22, *A*). From such cells the pyrenoid and eye-spot disappear. Within each such resting cell, four daughter cells arise, develop flagella and escape (Fig. 22, *C*).

A third form of resting cell occurs (Fig. 22, *A*). It is identical with that just described, except that it is colourless. Like its green counterpart it may divide to form four colourless daughter cells, which may be extruded from the mother cell-wall or divide yet further (Fig. 22, *B*, *C*, and *D*).

Again, paired resting cells occur. Two active green cells settle down together, become pressed against one another, and surround themselves with a common envelope. Such paired resting cells are either green or colourless.

The existence of paired and single colourless resting cells formed by the infecting organism in its free state completes the evidence as to the identity of this alga with the green cells of C. roscoffensis. For, as we have seen, though the alga may be ingested in its green, flagellated stage, the more usual mode of infection is by means of a colourless cell surrounded by a thick wall. This cell, lying in the central vacuole, undergoes division into daughter cells, which escape subsequently from the mother wall and are sown about the body of C. roscoffensis. The cell originally taken into the body may be paired or single. In the former case it gives rise to eight, in the latter case, to four daughter cells. We conclude, therefore, that the alga which is the infecting organism of C. roscoffensis, lives a double life. At times, it has the form of a green cell, at others, of a colourless cell. As a green cell it is holophytic, that is it manufactures photosynthetically its food materials. As a colourless cell it is a saprophyte, feeding like an animal on ready-made organic material. In its active stage, it is green and seeks the light : in its passive stage, it may be colourless and may live in darkness. Beneath the sand, therefore, where C. roscoffensis is born, abound vast numbers of the colourless infecting organism. On its emergence from the egg-membrane, the larva encounters them in plenty, lying and dividing on its egg-capsule and on any other organic debris. Should it escape infection—a rare contingency—C. roscoffensis may, as we have learned, return to the egg-capsule and thus incur it. Or, becoming positively phototropic, the larva moves up to the light. There, at the upper edges of the water-films, it finds assembled the green, flagellated organism. Beset in darkness and in light by the infecting organism, swallowing eagerly all the minute particles that come its way, C. roscoffensis cannot escape its destiny. A colourless or green cell is taken into the body and the plant-animal is formed.

So pleomorphic is the infecting organism that it occurs in yet other forms beside those described already. It may give rise by repeated divisions to groups of rounded cells lying together in a colony. Such a colonial form is known as a palmella stage and occurs in the life histories of various green algæ. It is remarkable that a colourless palmella form also exists. At any moment, a green member of the palmella may slip its mucilaginous coat and appear as a flagellated, active cell.

As to the name and position in the plant kingdom of the infecting organism we need say but little. Its characteristics are those of a group of primitive, green algæ know as the Chlamydomonadineæ. Like the infecting organism, the members of this group are unicellular, bear flagella of equal length and store their reserve food-material in the form of starch; they possess each an eye-spot, a pyrenoid and a cup-shaped chloroplast, enclosing a core of colourless, nucleated protoplasm. The only important respect in which the infecting organism differs from a typical chlamydomonadine cell is that, whereas the cell-wall of the latter consists of cellulose, that of the former has not given in our hands the reactions indicative of this substance.

Carteria, a genus of the Chlamydomonadineæ is characterised by the possession of four flagella, and so also is one species of Chlamydomonas (C. multifilis). Therefore the infecting organism should perhaps be referred to one or other of these genera. Or it may be that it belongs to a yet lower group. These are matters, however, for the systematist to decide.

What is certain is that the green cells of the body of C. roscoffensis once saw independent days, and that, for those cells, naked and deprived of nuclear material, these independent days are gone never to recur.

CHAPTER V

THE SIGNIFICANCE OF THE RELATION BETWEEN
COLOURED CELL- AND ANIMAL-CONSTITUENTS
OF THE PLANT-ANIMALS.

WE have discovered enough already, with respect
to the relations which obtain between algal cells and
animals in our composite plant-animals, C. roscoffensis
and C. paradoxa, to convince ourselves that the relation is not one of casual intimacy, lightly entered
upon and lightly abandoned; but one which is of
fundamental importance to the animals. Before it
was realised how vital was this association, we
entertained hopes of raising a colourless race of
C. roscoffensis, and, having done so, of comparing
the colourless with the green adult, with the purpose
of ascertaining what changes had arisen in the
organism as the result of the symbiosis. Though
this hope has not been fulfilled, though it has proved
a task, if not impossible, yet beyond our powers,
nevertheless the attempts to accomplish it have
brought to light facts which show how ingrained in the
lives of the worms is this habit of association with
the algal cells of their respective infecting organisms.

The attempts have also provided us with information with respect to the full significance of the association. In order to make clear the nature of this information, we will first consider very briefly the normal course of events in the development of C. roscoffensis in ordinary sea-water containing the infecting organism and then its behaviour in filtered sea-water which contains no infecting organisms.

Within a day or two of hatching, C. roscoffensis, maintained in ordinary sea-water, is found to have become infected. The colourless cell, the first sign of infection, divides and the colourless daughter cells are sown about the body. They become green, divide and re-divide till ultimately the many thousands of algal cells making up the green tissues of the organism, lie in dense masses in the body. For a time, the animal continues to feed : but, after a while, it abstains from ingesting solid food and lives a simple, easy life, fed by the products of the photosynthetic activity of its green cells. Later on, unsatisfied with the amount or kind of tribute which it thus receives, the animal begins to digest its algal cells and may continue the habit until the green tissue has, in larger measure, disappeared. In the meantime, however, C. roscoffensis has matured and produced eggs, the substance of which has been supplied by the green cells ; so, even though it now dies as the consequence of its ill-considered greediness, the continuance of the species is assured.

9—2

C. paradoxa exhibits a like behaviour. Though it never becomes a total abstainer from ingested, solid food, there are times when it makes inroads on its yellow-brown cells, and indeed, if supplies of solid food are withheld from it, C. paradoxa exhibits no less scruple than C. roscoffensis in raiding its algal cells.

But if C. roscoffensis is maintained in filtered sea-water, and hence prevented from becoming infected by its green algal associate, the course of events is very different. Even though diatoms and other micro-organisms are added to the filtered sea-water, C. roscoffensis, after a few days of active feeding, ceases from ingesting solid food-substances. For a while, the fat and other reserve food-substances contained in its body suffice for the needs of the animal, but, when these reserves are exhausted, starvation begins. In spite of any addition of food-material to the filtered sea-water—food-material which its infected green fellows are enjoying—uninfected C. roscoffensis abstains obstinately from ingesting it. It is waiting for the development of its green tissues which ought by this time to have been laid down in the body. Thus it waits, and starves, dwindles till it has become invisible to the eye, and, ultimately, after weeks of waiting, dies. If, before this happens, the algal infecting organism is added to the water, the animal may—should exhaustion be not too pronounced—ingest it. Having thus achieved infection, it is a changed being. Activity takes the place of lethargy and growth, of degeneration. In a few days it becomes, instead of a microscopic, transparent object, a visible, green organism.

The immediate problem is, how to explain this arbitrary behaviour of the uninfected organism. That it is at once a pathetic tribute to the dependence of C. roscoffensis on the infecting organism and a justification of the title of this book, is evident. Without the green cells, life to it is not worth living, and it dies though surrounded by a plentiful microflora of which in happier, infected circumstances it avails itself without stint.

Let us suppose that the tenor of normal development of an organism is not smooth and even, but abruptly intermittent ; that in the complex business of growing up—a business which involves many simultaneous processes and many processes which are necessarily consecutive—the consummation of one phase serves as the signal for the commencement of the next. Then, if, for one cause or another, one process does not complete itself, there will be no signal for the beginning of the consequent process. So, in respect of this series of processes, the organism never grows up. It exhibits the phenomenon of arrested development. The signal for full steam ahead with the next growth-process may be produced internally ; or it may be of external origin. The ivy,

which grows on the wall, awaits in vain for the signal
of high light intensity which is required to call forth
the development of its flowers.   Many plants exhibit
transient or permanent youth-forms ; that is, they
pass through or remain in juvenile stages of development.   A like phenomenon, with respect to the
organism as a whole, or with respect to single
organs is exhibited by animals and by man himself.
In the great number of cases, it is to be supposed
that the signal which is given or not given is of
internal origin.   It is very probable that such signals
for development are of a chemical nature.   The
important work of Starling (1906) has supplied
physiologists with a new method, capable of precise
application, in their work of analysing nervous
responses in animals and in plants.   Thus, he has
demonstrated that a secretion may be the result, not
of a nervous stimulus, but of the arrival at the
secreting organ of a definite chemical substance.
To give but one illustration of Starling's discoveries :
some time after food has been swallowed, the pancreas
begins to discharge pancreatic juice into the small
intestine.   Hence, the food, partially digested by the
stomach, is met, soon after its arrival in the small
intestines, by the pancreatic juice and acted on
in such a way that the materials it contains are
rendered soluble and diffusible and so capable
of passing into the blood-stream.   This purposeful,

automatic process of the production of pancreatic juice is independent of the nervous system. It occurs in the absence of all nervous connections between the intestines and pancreas. Now Starling has shown that the stimulus which induces secretion in the pancreas is due to a definite, chemical substance (secretin). This substance is produced in the small intestine as the result of the passage of food into that organ. It passes from the intestine into the blood-stream, is carried to the pancreas and gives the signal to that organ to commence its secretive activity. Such specialised, chemical stimulators Starling calls "hormones," and it is not to be doubted that they play an important part in inducing large numbers of normal processes which, as we know, arise as the consequences of antecedent processes. In plants in particular, it would seem that we must look to hormones, or chemical stimulators to provide us with an understanding of many phenomena which are at present ignored, or ascribed vaguely to nervous action. For example, the living tissue in the stem of plants, known as cambium, which is responsible, by continued growth and division, for the increase in thickness of the stem, occurs in young plants as definite, localised patches or sheets lying between the vascular-bundles. After the plant has reached a certain stage, the non-dividing cells of the cortex which are coterminous with the cambium cells, become, as it were, infected, and commence to divide. Then cells neighbouring these begin to divide and play the part of cambium, till finally a complete ring or hollow cylinder of actively dividing cells is formed in the stem ; and from this ring, which lasts as long as the plant lives, are produced new wood and new bast. Though no definite, chemical stimulator has been discovered in this case, we may feel sure that that is due to the fact that it has not been sought.

Applying the conception of chemical stimulators or hormones to the case of arrested development exhibited by C. roscoffensis, we may suppose that in this animal, the signal for the commencement of the later phases of development owes its origin to the presence, within the body of the animal, of the green algal cells, that, in the absence of these cells, the signal is not given, and that, consequently, development does not proceed.

Hence it would follow that no amount of feeding, either with diatoms or any other elements of the natural micro-flora and fauna existing in the environment of C. roscoffensis, can compensate for the lack of the hormone entrusted by custom with the task of signalling to the animal to proceed with the business of ordered development. On this view, the failure— which has been complete—to rear C. roscoffensis on artificial food, starch, sugar, peptone, protein, milk and prepared "human foods" of various kinds, is intelligible ; nor may we expect success to attend our attempts to raise a colourless race of C. roscoffensis till we have discovered the signalling substance produced by the green cells.

We turn now to another phenomenon exhibited by larval C. roscoffensis. Considered attentively, the rapid development of the green tissue in the infected animal is no less remarkable than the arrest of development in the uninfected animal. How comes it that an alien organism, intruding itself among the tissues of a young animal, is able to multiply so rapidly and extensively therein? It might be supposed that it was but a case of simple parasitism ; that the green cell lives and multiplies directly at the expense of the animal's cells. This, however, can scarcely be the case, for, in their early stages at least, the green cells keep themselves to themselves. They lie in vacuolar spaces out of direct contact with the animal cells. Hence any food-materials which they obtain from the body of the animal must be in a state of solution. Again, there is no evidence whatever that the green cells obtain access to any soluble food-substances which the animal has prepared for its own use. The time during which the increase of the green cells is greatest—soon after infection has taken place—is also the time when the animal itself is growing most rapidly. It is true that during this

period, the animal is ingesting solid food and therefore it is not impossible that the food-materials
obtained from this source may be shared alike by
the cells of the animal and the green, algal cells.
But, before we accept this view, we must enquire
into the conditions which obtain in the body of C.
roscoffensis at the time of infection, with the object
of ascertaining what sort of a "seed bed" for the
growth of the algal cells is provided by the body of
the animal.

The first fact which is brought to light by an
enquiry of this nature is that the association between
green cells and animal does not begin with the en-
trance of the green cell or its colourless antecedent
into the body. Before the relationship reaches this
condition of intimacy, animal and free green alga have
struck up an acquaintance based on the identity of their
mode of phototropistic response. Under the stimulus
of unilateral light, they both move in the direction
of the light and both proceed to the upmost edge of
the sea-water pools or streams in which they occur.
But this is not all. The free-living alga settles
down from time to time in the mucilage which forms
a slimy coating over the body of C. roscoffensis and,
withdrawing its flagella, passes into the condition of
a resting cell (Fig. 22). Inasmuch as the skin of the
animal provides the capsules which enclose the
clutches of eggs, it follows that, not infrequently, one

or more of the resting cells of the infecting organism
come to be included in the capsule-wall. Further,
apart from such chance inclusions, thanks to which
we were enabled to produce our pure cultures of the
alga, the egg-capsules appear to exert a definite
chemical attraction on the motile green cells. Thus,
if to a bulk of filtered sea-water containing egg-
capsules which have been laid under the cleanest
possible conditions, a number of the flagellated cells
are transferred, then, after a few hours, one or more
of the green cells will be found to have settled down
on each capsule. Yet more striking results are
obtained if a capsule is suspended in a hanging drop
—that is, a drop of sea-water which depends from
the under side of a microscope cover-glass—and if
a number of the flagellated cells are added to the
drop. On observing such a preparation under the
microscope, the motile green cells are seen to approach
the capsule, to swarm about it, to press in close ranks
into the soft, gelatinous wall and so embed themselves
in the envelope. We conclude, therefore, that the
egg-capsule exercises an attractive (chemotactic)
influence on the flagellated algal cells ; or, in other
words, that a definite substance diffusing out from
the capsule-walls induces a tropistic (tactic) movement in the motile, algal cells of such a nature that
they approach the source whence the chemical substance emanates. The behaviour of the green cells,

which thus settle on and in the capsule, proves that they find in it a favourable medium for growth. Within a few hours, each green cell, having withdrawn its flagella, increases considerably in size and, whilst retaining its green colour, takes on a granular appearance. The eye-spot and pyrenoid become fainter and the cell undergoes division. In the daughter cells thus produced, a series of successive divisions occur till a loose colony of green cells is formed—such a colony, in short, as that which enabled us to determine the nature of the infecting organism (p. 120). In egg-capsules, some of the eggs of which have died, the green cells find yet richer supplies of food-material and increase the more rapidly. These observations give us a hint as to the nature of the food-materials contained in the capsules, which serve for the rapid increase in the green cells. For though, as we have learned, green plants have at their command unlimited supplies of the raw materials, carbon-dioxide and water, for the manufacture of carbohydrates, they are by no means in so happy a situation with respect to the raw materials for the synthesis of organic, nitrogen-containing compounds. A green plant growing with its roots in the soil, and relying on inorganic salts—nitrate of potash, etc.—for its supplies of nitrogen-containing, raw material, is often hard put to it to obtain enough of these nitrogen compounds wherefrom to manufacture its proteins, and thus to augment its living substance, integral parts of which consist of organic, nitrogen-containing compounds. That such plants suffer not infrequently from nitrogen-hunger is one of the most important agricultural discoveries of the last century. As a consequence of the recognition of this fact, many thousands of tons of nitrate of soda from the nitrate beds of S. America and equally vast quantities of sulphate of ammonia—a bye-product of the distillation of coal—are added annually by the farmer to his land. Nor is the origin of this nitrogen-deficit far to seek. The nitrogen contained in the nitrates of the soil comes in the plant to form a constituent of the organic nitrogen compounds, such as the proteins. The plant dies and decays, or is eaten and the eater decays. Ultimately, as the result of these processes of decay, water and carbon-dioxide are liberated and may at once be brought again, by the agency of the green plant, into the vital circulation. Synthesised to form carbohydrates, these substances are once more available for the nutrition of plants and animals. But with respect to nitrogen it is otherwise. The organic nitrogen compounds of the dead animal or plant are broken down by the bacterial and fungous agents of decay into a series of simpler forms which, acted on by yet other of the ordered army of saprophytic micro-organisms, yield finally ammonia and nitrogen. The nitrogen leaks away into the atmosphere and contributes to the 79 per cent. of nitrogen gas which is contained in the air. The ammonia may leak away also—as every dung-hill testifies—or it may be fixed in the soil by the agency of certain nitrifying micro-organisms. These bacteria convert the ammonia into nitrates and the nitrates so formed become available to the roots of the green plant. On the other hand, the nitrates of the soil may be seized upon by yet other, denitrifying micro-organisms and, becoming converted into ammonia compounds, may be lost to the vital circulation. The constant leakage of nitrogen from combined forms to the free and inert form of nitrogen gas results in a shortage of nitrogen available for the formation of the nitrogenous food of plants. We may thus speak of the problem which besets all living organisms—that of obtaining adequate supplies of organic nitrogen compounds — as the nitrogen problem, and we may well believe that the sum-total of life supported on our planet is determined ultimately by the amount of available nitrogen present in the earth and sea. Occasionally, organisms are met with which have solved the nitrogen problem in a fundamentally satisfactory manner. Among such organisms are nitrogen-fixing bacteria, leguminous plants and man. Each of these organisms has evolved methods of bringing back into vital circulation the nitrogen which has escaped as nitrogen gas into the air.

The nitrogen-fixing bacteria which occur in the

soil and also in the sea, possess the power of
causing free nitrogen to enter into combination with
other elements and so to serve as material for the
construction of the vitally necessary proteins. The
leguminous plants, clovers, peas, lupins, etc., do it—
or rather get it done for them—by entering into
association with a certain species of nitrogen-
fixing micro-organism. This organism, Pseudomonas
radicicola, enters the root and increases in its
tissues. Under the stimulus of this micro-organism,
the root swells locally to form nodules or tubercles.
Later, when the nodule-organism has accumulated
considerable quantities of organic, nitrogen compounds, the tissues of the root destroy it, raid its
stores and, living on the nitrogen-plunder, are able,
unlike other plants, to grow in soils which are
deficient or even lacking in inorganic, nitrogen compounds. Thus, the gorse occupies vast tracts of
sterile, sandy wastes in Brittany and elsewhere, and
the traveller in spring may journey for miles between
tree-like groves of gorse ablaze with golden blossom,
every particle of which owes its presence in the air
to the nitrogen-fixing bacteria at work in the roots
underground. These bacteria it is which have provided
the essential, organic nitrogen compounds without
which the tissues of the flowers could not have been
formed. Large tracts of waste land in Germany,
America and other parts of the world have been

rendered amenable to cultivation by planting with lupins. The roots of these plants, beset with nodules, decay in the ground, release nitrogen-compounds, hitherto deficient in the soil, and thus, by their decay, admit of the growth of plants which rely entirely on "fixed" or combined nitrogen. It is computed by competent authorities that in Germany alone no less than 500 million pounds of nitrogen are secured annually from the air through the activity of the root-tubercle bacteria associated with leguminous crops.

It is a grim commentary on the mode and rate of progress of agricultural science that these discoveries of the men of science yesterday were among the accepted commonplaces of the ancients. Thus Pliny observes that "the bean ranks first among the legumes and it fertilizes the ground in which it has been sown as well as any manure."

Man solves the nitrogen problem by including legumes in his crop-rotations, by transporting nitrates from Chili to his European fields and—more recently—by effecting a combination of the nitrogen of the air with oxygen or other elements, utilising for this purpose electrical energy. Where water-power is available for the generation of electricity, factories, destined to play an increasingly important part in the solution of the nitrogen problem, are at present at work turning out large quantities of calcium nitrate or other nitrogen-containing compounds.

These compounds, put into the soil, are each a source whence the green plant may obtain the raw materials for the synthesis of organic nitrogen and thus increase the supplies of material essential for the development of brain and muscle in animals and man.

The fact of nitrogen-hunger is, then, no small matter of mere academic importance. It touches the future of man himself and presents a problem which every living organism must solve. The supply of available nitrogen is a limiting factor of life. Let us see what bearings the fact of nitrogen-hunger have on the economy of C. roscoffensis and C. paradoxa.

That nitrogen-hunger presses as hardly on marine organisms as on those which live on the land is undoubted. Recent investigations have shown that the amount of combined nitrogen present in sea-water, in a form available to plants for synthetic purposes, is extremely low. Thus, according to Johnstone (1907), the amount of nitrogen compounds in Baltic and North Sea water may be taken as about ·2 millegrams (= ·003 grains) in a litre, or about two parts in a million. No wonder that marine animals are always hungry! No wonder either that the free, flagellated infecting organism of C. roscoffensis settles down on the egg-capsules to avail itself of any crumbs of nitrogen compounds that it may find there. Nor is it remarkable that, finding a certain amount of combined nitrogen, it begins to divide and soon forms a colony of numerous green cells.

Now, as we have indicated previously, C. roscoffensis and C. paradoxa are remarkable among the Turbellarian worms in possessing no excretory system. Unlike their allies, they possess no apparatus for the systematic discharge of the waste products of their metabolism. Hence such products, compounds of nitrogen of a kind useless to the animal, are stored in the tissues of the body. But such substances, though useless for the nutrition of the animal, serve well for plants. Even a terrestrial green plant is very catholic with respect to the compounds of nitrogen which it takes up and utilises for the synthesis of proteins. Thus, experiment has shown that the root-system of a green flowering plant is capable of absorbing, not only nitrates and, in many cases, ammonium salts, but also such organic, nitrogen-containing substances as urea, uric acid, asparagine and many others. Now the infecting organism of C. roscoffensis occurs, as we know, in a colourless as well as in a green stage, and, in the colourless form, it can obtain its food materials only after the manner of an animal, that is, in combined organic form. So that its powers of taking up and utilising organic nitrogen compounds are likely to be even more marked than those of a self-supporting green plant. This con-

jecture is confirmed by experiment. Comparative
cultures of the free stage of the infecting organism
have demonstrated that the alga flourishes better when
supplied with nitrogen in the form of uric acid than
when it is supplied with a nitrate (potassium nitrate).

Thus our argument brings us to the following
position : We have evidence that the infecting
organism increases rapidly as soon as it gains access
to the body of the plant-animal. We know that it
is able to utilise organic nitrogen compounds such
as uric acid for the construction of its proteins. We
know, further, that no apparatus for the removal of
waste nitrogen compounds, uric acid, urea, etc., occurs
in the bodies of C. roscoffensis or C. paradoxa. The
conclusion forces itself upon us that the green and
yellow-brown cells in the bodies of their respective
hosts obtain access to and utilise the stores of waste
nitrogen-compounds accumulated therein. Or, to
put the same idea in another way, green cells and
yellow-brown cells constitute the excretory organs
of C. roscoffensis and of C. paradoxa respectively.
The plants flourish in the bodies of these animals
because there they discover large accumulations of
waste nitrogen compounds: the animals, looking to
the algæ to come and take charge of the work of
getting rid of these waste substances, have ceased
to construct any excretory apparatus, whatever.
Hence it is not surprising that, when the algæ fail

to appear in their bodies, the animals suffer. It may be that the death of uninfected animals is not merely the consequence of starvation, but is at all events hastened by poisoning due to the accumulation in the tissues of the products of nitrogenous metabolism. According to this view, uninfected C. roscoffensis dies as the consequence of an aggravated attack of " uric acid trouble."

Evidence is not lacking in support of this somewhat fantastic suggestion. Thus, if larval C. roscoffensis are protected from infection and kept without food, as their large store of reserve food-material derived from that contained in the eggs, disappears, numerous vacuoles charged with long, acicular, crystalline bodies make their appearance in the tissues. The vacuoles and crystals increase in numbers till they present a most striking appearance. These crystals represent, in all probability, the waste products of nitrogen-metabolism.

Now, in infected animals, the crystalline bodies do not occur, and if animals in which they are present are caused to become infected by the green algal cells, the crystals disappear as fast as the green cells develop. Whence we may infer that the materials of which the crystalline bodies consist are used for the nutrition of the green cells.

The evidence which C. roscoffensis provides in favour of our hypothesis is, of course, but slender.

Let us appeal therefore to C. paradoxa. A far more greedy feeder than the green species, its accumulations of nitrogenous waste substances are much larger than are those of its ally. Inspection of the Frontispiece or of Fig. 4 shows well-marked, granular bands across the body of the animal. These bands consist probably, as von Graff has suggested, of urates. They are slight in the young animal, increase as it matures, but may disappear as the period of egg-laying arrives, at which time the yellow-brown cells have developed to their full extent.

In order to establish our hypothesis we must demonstrate that the yellow-brown cells of C. paradoxa actually make use of such substances—presumably uric acid or urates—as are stored in the body.

For this purpose, two modes of experimentation were adopted. In the first method, batches of animals of similar sizes and origin were maintained in the light in filtered sea-water to which uric acid was added and their condition was compared with that of animals kept in filtered sea-water containing no uric acid. Preliminary observations showed that the uric acid added to the sea-water was taken up readily by the animals and stored in vacuoles in the tissue of the digestive tract. Examination and measurement of animals from the two batches—those in filtered sea-water only and in filtered sea-water plus uric acid—proved that the latter were, after twenty-one days, considerably larger than the former.

The experiment was continued. During the following weeks the animals in filtered water, dwindled, lost all their yellow-brown cells, became of microscopic size and died. On the other hand, after upwards of thirteen weeks, specimens of the animals in filtered water plus uric acid were alive, of a recognisably brown colour and possessed of many normal, yellow brown cells.

We thus have proof that when C. paradoxa is kept in the light, so that its yellow-brown cells may photosynthesise, and when uric acid is supplied, this substance serves as a source of nitrogen to the yellow-brown cells. Moreover, in these circumstances, the materials manufactured by the yellow-brown cells serve not only for the nutrition of the alga but also for that of the animal. This, however, means that the yellow-brown cells contribute not only fatty but also nitrogenous, protein-forming material to the animal. That this is the case the results of the second mode of experimentation render highly probable.

Here, in lieu of determining the effect of uric acid on the life of algal cell and animal, its influence on egg-laying was investigated. The experiment consisted in maintaining equal numbers of similar animals in filtered sea-water, under conditions which were identical except for the fact that one lot received uric acid. The animals supplied with no extra nitrogen laid nine clutches of eggs, whereas the animals supplied with extra nitrogen laid twenty-seven clutches.

The results of the two sets of experiments just described serve to account for the rich development of algal cells within the bodies of the plant-animals. In their free state, these algæ, like all marine plants, run grave and frequent risk of nitrogen-starvation, or at all events of having their increase limited by the shortage of available nitrogen in the sea. Wherever there is any leakage of nitrogen compounds—and traces of combined nitrogen must be given off from such animals of C. roscoffensis and C. paradoxa— marine, motile plants will congregate. Congregating about our plant-animals, such minute organisms are ingested indifferently. Out of this mixed infection C. roscoffensis and C. paradoxa make each a pure culture, the one of green cells the other of yellow-brown cells. Established in the body, the algal cells find themselves transferred from a region of scarcity to a land of plenty. Outside, in the open sea, the amount of nitrogen available is but small and the claimants for a share of it innumerable: within the body, the amount of suitable, combined nitrogen is large and at the exclusive disposal of the algal visitors. In such Capuan circumstances, the algal cells grow and divide luxuriantly. Their photosynthetic activities

increase, for only in the presence of plentiful supplies of nitrogen does the chlorophyll-apparatus work
well. Large quantities of carbohydrate material
are produced in the algal cells—enough for the needs
of these cells and also for those of the animal. All
goes well, so well indeed that C. roscoffensis, less
conservative than its ally, contents itself entirely
with the supplies of food-material, of fat and also
of organic nitrogen compounds, provided by its green
cells and abandons the practice of fending for itself.
The ample tribute which it receives suffices for its
needs and also for the provision of its eggs. But the
weakness of the system here discloses itself. This
handing of nitrogen-containing substances to and fro
from animal to plant and from plant again to animal
cannot go on indefinitely or without loss. Sooner or
later, the animal finds itself lacking in essential,
nitrogen-containing food-materials. Supply fails to
equal the demand. Then the animal is under the dire
necessity of digesting its algal cells. To satisfy an
imperious, present need, the plant-animal destroys
the source of its supplies.

Thus the animal repudiates the association and,
having digested its green cells, C. roscoffensis dies
of the very complaint—nitrogen-hunger—which the
green cells sought to avoid by their intrusion into
the body of the animal. To dismiss the association
between animal- and plant-constituent of the plant-

animals by labelling it symbiosis is to miss the varying significance of the association. Looking at the
relationship from the standpoint of the animal, it is
one of obligate parasitism. Apart from their algal
cells, C. roscoffensis and C. paradoxa are unable to
live. The existence of either species depends upon
the infection of the individuals of each successive
generation. Where the infecting organism is absent,
there C. roscoffensis does not exist. Hence its restricted range. From the standpoint of the species,
"infecting organism," the relation of certain of its
individuals with C. roscoffensis or C. paradoxa is an
episode without significance. Unlike the animal,
which bears the inherited impress of the relation
in lack of excretory system and in the habit of
patient waiting—abiding the question of infection—
the alga is free. Of a swarm of flagellated green
cells, some small percentage meet the picturesque
fate of forming a tissue in the body of an animal.
The others pursue a less romantic adventure, either
as green, self-supporting organisms or as colourless
cells which batten on the offal of the sea.

From the standpoint of the ingested algal cell,
association with the animal means a successful solution of the nitrogen problem. It sacrifices its
independence for a life of plenty. This universal
nitrogen-hunger is a misery which makes strange
bed-fellows.

It is noteworthy that the interpretation, in terms of the hypothesis of nitrogen-hunger, of the relation between animal and algal cell throws light on the facts, already referred to, concerning the distribution of algal cells in various marine animals. Analyses have demonstrated (Johnstone, 1907) that the amount of combined nitrogen present in sea-water is less during the warm months (e.g. August) than during the cold months of the year, and that it is less in the warmer seas (Mediterranean) than in the colder seas (Baltic and North Sea). Now, as we have mentioned, certain animals possess green or brown algal cells in one part of their range of distribution but lack them in other parts. Thus Noctiluca, colourless in the North Atlantic, is green in the Indian Ocean. Whence it would appear to follow that where the stress of nitrogen-hunger is more acute, there the association between algal cells and animals manifests itself.

One word more and one more speculation and our work is done. The colourless phase in the life-history of the infecting organism of C. roscoffensis, the colourless state of the just-ingested algal cells both in C. roscoffensis and C. paradoxa, and the rapid assumption of their proper pigments by the infecting cells after they are established in their respective animal quarters suggest that the colourless phase is itself the outcome of nitrogen hunger. Such colourless phases are known to occur in the life histories of other micro-organisms, in diatoms, in various species of Chlamydomonas and in Flagellates (Euglena), and it is stated generally that they may be induced by increasing the amount of soluble carbohydrate in the culture medium. But in the cases of the algal-infecting organisms of our plant-animals, the rapid development of the chlorophyllous pigment appears to be associated with the increase in the amount of available nitrogen. So that, if this is the case, the colourless phase would appear to be brought about, not by excess of food-material, but by lack of nitrogen. It may well prove to be that the colourless sapro-phytic phases exhibited by such organisms as those just mentioned—diatoms, etc.—are each a symptom of nitrogen-hunger. For, failing proper supplies of nitrogen compounds, no amount of carbohydrate photo-synthesis will keep the organism from starvation. Indeed, the more the carbohydrate photosynthesis, in-volving as it must the wearing out and reconstruction of the nitrogen-containing chlorophyll machinery, the acuter will be the nitrogen-hunger; whereas, on the contrary, a shutting down of the photosynthetic process will effect economies in the use of organic nitrogen compounds and thus postpone the evil day of nitrogen-starvation. Though the facts are not yet available for a confident statement, the hypothesis may be proposed that saprophytism generally depends for

its inception on nitrogen-hunger. It is tempting
to push this hypothesis to its limits, and to imagine
that the great saprophytic groups of the fungi
and the bacteria owe their origin to the changed
mode of nutrition imposed upon them by lack of
nitrogen. That the fungi are examples of descent
by reduction is undisputed. All the evidence points
to their derivation from chlorophyll-containing algal
ancestors. Having lost their chlorophyll, and, with
it, their powers of photosynthesis, they are now con-
demned to obtain both carbon and nitrogen in the
form of organic compounds and hence are compelled,
with the bacteria, to play the part of Nature's
scavengers. In their quest for food, they settle
either on the dead remains of plants or animals,
or, invading the living organism, they exchange a
saprophytic for a parasitic mode of life.

The hypothesis suggested here is that the first
and fatal step from independence to dependence was
the outcome of the nitrogen scarcity which exists
in Nature. Confronted with inadequate supplies of
nitrogen, the photosynthetic activity of their chlorophyll apparatus was brought to a standstill. The
organisms, unable to obtain supplies of inorganic
nitrogen compounds, were constrained to resume their
powers, never wholly lost, of absorbing nitrogen compounds in organic form. But such organic nitrogen-
containing compounds contain also carbon. Hence

supplies of this element were obtained together with nitrogen. In these circumstances, the expensive chlorophyll apparatus ceased to be worth its upkeep and, wearing out, proved to be too costly in nitrogen to be replaced. Thus the organism, now devoid of chlorophyll, was reduced to a condition in which it obtains directly from its environment as much carbon in combined form as is of use to it and as much combined nitrogen as it can get. It has become a saprophyte.

Should this hypothesis of the origin of saprophytism be established, C. roscoffensis and C. paradoxa will rank high in interest among organisms as suggesting the route along which far-reaching evolution has travelled. In any case, it may be claimed for our plant-animals that they have anticipated the advice of Candide and live to cultivate their gardens.

Both C. roscoffensis and C. paradoxa possess self-sown, well-tended, highly productive gardens, and if they could but learn how to bequeath packets of vegetable seed to their descendants, they might lose their animal characteristics altogether and become, C. roscoffensis a green plant, and C. paradoxa a yellow-brown plant. As it is, the garden has to be replanted in the individuals of the successive generations and so they remain plant-animals.

BIBLIOGRAPHY·

For more complete lists of the literature dealing with the subject of symbiosis between animals and plants see the Bibliographies attached to the memoirs published by Messrs Gamble and Keeble in the *Quarterly Journal of Microscopic Science* (1903, 1907, 1908).

1879. Geddes, P. Observations on the Physiology and Histology of Convoluta Schultzii. Proc. Roy. Soc. XXVIII. pp. 449—457.

1880. Darwin, C. and F. The Movements of Plants, p. 523.

1898. Williams, J. Lloyd. Reproduction in Dictyota dichotoma. Ann. of Bot. XII. pp. 559—560, 1898; and The Periodicity of the sexual cells in Dictyota dichotoma. Ann. of Bot. XIX. pp. 531—560. 1905.

1900. Goebel. Organography of Plants. Eng. Trans. Univ. Press, Oxford. p. 244.

1903. Bohn, G. Sur les mouvements oscillatoires des Convoluta roscoffensis. C. R. Ac. Sc. Oct. 1903.

1903. Gamble, F. W. and Keeble, F. The Binomics of Convoluta roscoffensis. Q. J. M. S. LVII. 1903.

1904. Semon, R. Die Mneme. W. Engelmann. Leipzig, 1904.

1906. E. H. Starling. Recent Advances in the Physiology of Digestion. London, 1906.

1907. Johnstone. The Law of the Minimum in the Sea. Sci. Progress, II. No. 6. Oct. 1907 ; and Life in the Sea. Univ. Press, Cambridge. Biological Series.

1907.   Keeble, F. and Gamble, F. W.   The Origin and Nature of the Green Cells of Convoluta roscoffensis.   Q. J. M. S. LI. Part 2.   1907.

1908.   Keeble, F.   The Yellow-brown cells of Convoluta paradoxa. Q. J. M. S. LII.   Part 4.   1908.

1909.   Loeb, J.   Experimental study of the influence of Environment on Animals.   Essay in Darwin and Modern Science. Univ. Press, Cambridge.

INDEX

Alcyonium (British), 101
Alcyonium ceylonicum, 101

Bohn, G., 64
Butler, Samuel, 49

Carteria sp., 129
Chemical stimulators (hormones), 135
Chemotactism, 139
Chlamydomonadineæ, 128
Chlamydomonas, 129
Chlorophyll, 87
Chloroplast, 85, 86, 105
Convoluta paradoxa :
  Background, influence of, 45, 48, 50
  Behaviour in constant darkness, 65
  Bristles, 8
  Cilia, 7
  Digestion of green cells by, 83, 97
  Digestive system, 11
  Eggs, 13; periodicity of production of, 24
  Egg-laying, conditioned by illumination, 31
  Eyes, 9, 54
  Fat, 89
  Feeding habits, 81, 83, 97, 98

Convoluta paradoxa (cont.):
  General aspect, 5
  Glands (pigmented), 9, 54
  Gravi-perception, 10
  Gullet, 11
  Habitat, 7, 19
  Mouth, 11
  Otocyst, 9
  Paradoxa zone, 17
  Periodicity of egg-laying, 24, 34
  Phototropism, 42
  Secretion of fat by yellow-brown cells of, 92
  Starvation, resistance to, 94
  Tidal migration, 21
  Tropistic response to light, 42
  Uric acid, effects on egg-laying, 150
  Vacuoles, 11
  Yellow-brown cells of, 75, 84, 91, 95, 147
Convoluta roscoffensis :
  Background, influence of, 45, 48, 50
  Chlorophyll in, 87
  Cilia, 7
  Dark-rigor, 60
  Digestion of yellow-brown cells by, 95

Convoluta roscoffensis (*cont.*) :
Digestive system, 11
Eggs and egg-capsules, 13, 56 ; periodicity of laying of, 26
Excretory organs (absence of), 147
Eyes, 9
Feeding habits of, 77, 81, 97
General aspect, 5
Gravi-perception, 10, 39
Green cells of, 75, 84, 105; algal nature of, 118; life history of, 132; origin, 108
Gullet, 11
Habitat, 7, 14, 18
Light-rigor, 60
Mouth, 11
Nucleus, 110
Nuclear degeneration, 112
Otocyst, 9
Periodicity of ascent, 62; of egg-laying, 26
Photosynthesis by the green cells of, 81, 87
Phototonic effect of stimulation, 59
Phototonus, 59
Phototropism, 41, 52
Reaction to monochromatic light, 54
Regeneration of, 27
Response to vibration, 15
Rhythmic ascent and descent, 19, 62
Roscoffensis zone, 17
Simultaneous stimuli, 44, 45
Size, 16
Starch in green cells, 87
Starvation, resistance to, 94
Tidal rhythm, 63, 67
Tonic influence of light, 67
Tropistic response to light, 41

Convoluta roscoffensis (*cont.*) :
Vacuoles, 11
Vibration, response to, 62, 67
Copepods, tropism of, 69

Dictyota dichotoma, 34
Directive stimuli, 40

Echinocardium sp., 100
Elysia sp., 100
Eudendrium racemosum, 33
Euglena viridis, 103
Eye-spot, 102, 110

Flagella, 124
Fungi, 156

Geddes, P., 87
Goebel, K., 33
Gravi-perception (by roots), 38
Green cells of animals, 82, 100
Green cells of C. roscoffensis, algal nature of, 128; colourless phase of, 126; cultivation of, 115; structure and life history, 110, 123
Green light, and marine organisms, 54

Haberlandt, G., 105
Hering, Prof. E., 49
Hippolyte varians, 47
Hormones, 135
Hydra viridis, 100

Ivy, 32

Lankester, Sir Ray, 114
Leucoplast, 105, 109
Lichens, 107

Light, influence of, on plants, 32; on regeneration of polyps, 33
Loeb, J., 29, 69

Macrocytes, 124
Microcytes, 124
Mneme (memory hypothesis), 49
Monochromatic light, 54
Mysis sp., 47

Nervous impulses, 39
Nitrogen-compounds in sea-water, 145, 154
Nitrogen-fixing bacteria, 143
Nitrogen-hunger, 145
Nitrogen-problem, 142
Noctiluca, coloured cells of, 101

Otocyst, 9

Palmella, 128
Photosynthesis by green plants, 78; by C. roscoffensis, 87
Phototropism of Copepods, 69
Prawns, 45
Protoplast (cell), 79
Pseudomonas radicicola, 143
Pyrenoid, 85

Reflex action, 43
Reflex arcs, 40
Reproduction, periodicity of, in brown sea-weeds, 84
Rhizobium leguminosarum (= Pseudomonas radicicola), 143
Roscoffensis zone, 17

Salamandra atra, 29
S. maculosa, 29
Saprophytism, origin of, 157
Schimper, A. F. W., 114
Secretion, 135
Semon, R., 48
Simultaneous stimulation, 54
Starch, 87
Starling, E. H., 134
Starvation, 94
Symbiosis, 106, 143, 153

Tactic response to stimulation, 41
Tonic effect of light-stimulation, 58, 67
Tropistic response to stimulation, 41, 69

Unconscious memory, 49
Uric acid, absorption of, by C. paradoxa, 149

Vacuoles (digestive), 11
Von Graff, 149
Vöchting, H., 33

Williams, J. L., 34

Yellow-brown cells of animals, 82, 100; of C. paradoxa, 75, 84, 91, 95, 147

Zoobothrium sp., 100
Zoochlorella, 101, 105
Zooxanthella, 101

For EU product safety concerns, contact us at Calle de José Abascal, 56–1°,
28003 Madrid, Spain or eugpsr@cambridge.org.

www.ingramcontent.com/pod-product-compliance
Ingram Content Group UK Ltd.
Pitfield, Milton Keynes, MK11 3LW, UK
UKHW010850090126
466816UK00011B/145